Ulrich Müller-Kolck

The Logical Structure of Clinical Medicine

To Peter Nissen

The Logical Structure of
Clinical Medicine

Ulrich Müller-Kolck

Bibliographische Information der Deutschen Bibliothek
Die Deutsche Bibliothek verzeichnet diese Publikation
in der Deutschen Nationalbibliographie; detaillierte bibliographische Daten
sind im Internet über http://dnb.ddb.de abrufbar.

Ulrich Müller-Kolck
The Logical Structure of Clinical Medicine
Norderstedt: Books on Demand 2010

ISBN 978-3-83915-419-9

1. Auflage 2010

© 2010 by Books on Demand GmbH
In de Tarpen 42, 22848 Norderstedt
Alle Rechte vorbehalten
Herstellung und Verlag: BoD - Books on Demand, Norderstedt
Gedruckt auf alterungsbeständigem Papier
Printed in Germany

www.bod.de

Contents

PREFACE 9

1 INTRODUCTION 11

2 METHOD 13

3 PATIENT IMAGES 15

3.1 PATIENT SELF-IMAGE (PSI) 16
3.2 DOCTOR SELF-IMAGE (DSI) 21
3.3 MEDICAL DESCRIPTIONS 22
3.4 NOSOLOGY ENTITY (NE) AND NOSOLOGY NETS (NN) 24
3.5 PATIENT IMAGE (PI) 29
3.6 PARTIAL POTENTIAL PATIENT IMAGE (PPPI) 29
3.7 POTENTIAL PATIENT IMAGE (PPI) 31

4 DISEASE COURSE 35

5 HEURISTIC PROCESSING 43

5.1 HISTORY BASED-LEARNING 43
5.2 DIAGNOSTIC INFERENCING 48
5.3 PROGNOSTIC PLANING 52
5.4 ARGUMENTATION SPACE 57

6. SUMMARY 61

7. EXERCISES: TOOLS FOR COGNITIVE PROCESSING 65

8 APPENDIX 69

8.1 SETS 69
8.2 RELATIONS AND FUNCTIONS 70
8.3 DEFINITIONS 71
8.4 ALGORITHMS 74

9 REFERENCES 77

Preface

The text is the core of a knowledge-engeneering lecture in computer science (AI-working group, Technical Informatics, Universitiy Bielefeld, 1992-95) introducing master students to basic reasoning principles in clinical medicine. The very general approach to the practice of medicine oversimplifies the complexity of real clinical settings. Many of the issues would require extended discussions. The text provides a proto-theory.

Acknowledgements

The author is much in dept to Reinhard Flessner for reading of the text and his criticisms.

1 Introduction

In our professional medical language the term "clinical medicine" denotes all the activities of solving individual health problems under real world conditions. Clinical medicine has often been analysed under the focus of processing a nosological taxonomy for the classification of individual disease cases. From the viewpoint of medical practice cognitive processing is more than solving classification tasks.[1] In contrast the core structure of clinical reasoning is the dynamic concept of the individual disease course. Physicians use basic clinical heuristics to reconstruct idiosyncratic disease courses. Both elements, disease course and reasoning heuristics, are necessary to apply medical knowledge to the real world problems of the patients.

My thesis is that the practice of medicine basically depends on three general concepts: an individual patient image, the individual disease course of this patient, and an expectation of the future development of the disease. I analyze the logical structures of these concepts in chapter 3 (patient image) and chapter 4 (disease course).

Physicians use heuristics for the cognitive processing of these concepts under real world conditions. Heuristics have the advantage that they guide medical decisions within fuzzy boundaries. Heuristics are open for new information, for intuitions.[2] In chapter 5 basic clinical heuristics are condensed into three algorithms: a history-based-learning algorithm, a diagnostic-inferencing algorithm, and a prognostic-planning algorithm. Chapter 6 summarizes the results of my analysis.

---

[1] Braunwald E, Isselbacher KJ, Petersdorf RG, Wilson JD, Martin JB, Fauci AS (eds.). Harrison's Principles of Internal Medicine, New York, et al.: McGraw-Hill Book Company, 11th ed. 1987. Hurst, J.W. (ed.), Medicine for the Practicing Physician, 2nd edition, Boston et. al.: Butterworth, 1988. Kahneman D, Slovic P, Tversky A (eds.). Judgement under Uncertainty: Heuristics and Biases. Cambridge, England: Cambridge University Press 1982. Macleod John (ed.). Clinical Examination. Edingburgh, London, NewYork: Churchll Livingstone, 1979. Murphy EA. Probability in Medicine. Baltimore, London: The Johns Hopkins University Press, 1979. Murphy EA. Skepsis, Dogma and Belief. Uses and Abuses in Medicine. Baltimore, London: The Johns Hopkins University Press, 1981.

[2] Gigerenzer G. Gut Feelings. New York: Viking. 2007. Gigerenzer G. Fast and Frugal heuristics: tools of bounded rationality, in: Koehler D, Harvey N (eds.): Blackwell handbook of judgement and decision making. Oxford 2004, 62-88. Schwartz S, Griffin T. Medical Thinking. The Psychology of Medical Judgment and Decision Making, Springer: New York, 1986.

Although computer-based medical information processing has tremendously progressed in the last years, there is no convincing penetration of these technologies into our everyday practice of medicine. Chapter 7 enumerates possible applications of the proposed theory. A list of reasonable exercises may encourage cognitive engineers to design tools to support cognitive processing of individual patient data. These tools should realize the proposed algorithms. They should be judgment assistants or "tools for thinking"[3] for the practicing physician.

---

[3] Hurst, J.W. (ed.), a.a.O., p13.

2 Method

The author was interviewed as a medical expert during the knowledge acquisition process for the HYPERCON-project (HYPERCON 1992 – 1995: University of Bielefeld, Faculty of Technology, Knowledge Based Systems - Artificial Intelligence Research Group, Head: Ipke Wachsmuth).[4] The reconstruction of general structures in clinical medicine is based upon the transcripts of these interviews.

The formal approach for the reconstruction uses Patrick Suppes's method of axiomatizing theories by defining set theoretical predicates.[5] His informal set theoretic approach has later been extended to the so called "Structuralist View of Scientific Theories".[6] Following these ideas the basic structures in

---

[4] Heller B, Meyer-Fujara J, Schlegelmilch S, Wachsmuth I. HYPERCON: ein Konsultationssystem zur Hypertonie auf der Basis modular organisierter Wissensbestände. University of Bielefeld ; Fakultäten. Technische Fakultät, in: G. Barth et al.: Anwendungen der künstlichen Intelligenz: KI-94, 18. Fachtagung für Künstliche Intelligenz, Saarbrücken, 22./23. September 1994, Berlin: Springer: 1994, 155-169. Heller B. Modularisierung und Fokussierung erweiterbarer komplexer Wissensbasen auf der Basis von Kompetenzeinheiten. Bielefeld, Univ., Diss., 1995. Sankt Augustin: Reihe: Dissertationen zur künstlichen Intelligenz. 1996. Heller B, Herre H, Lippoldt K, Loeffler M. Standardized Terminology for Clinical Trial Protocols Based on Top-Level Ontological Categories. In: Kaiser K et al (eds.) Computer-based Support for Clinical Guidelines and Protocols, IOS Press 2004, 46-60. Meyer-Fujara J, Heller B, Schlegelmilch S, Wachsmuth I. Knowledge-level modularization of a complex knowledge base, in: Bernhard Nebel et al. Advances in artificial intelligence: proceedings, eds. (Lecture notes in computer science; 861), Berlin: Springer, 1994, 214-225. Müller U. SCREEKON: Medizinisches Konsultations-system für Screeningfunktionen am Beispiel des therapeutischen Managements von Patienten mit cerebralen Durchblutungsstörungen, Angewandte Inforamtik 2/1989, 76. Müller-Kolck U. Medizinische Therapieentscheidungen mit SCREECON, in: Savory SE (ed.) Expertensysteme: Nutzen für ihr Unternehmen. Ein Leitfaden für Entscheidungsträger. München, Wien: R. Oldenbourg; 1989. Müller-Kolck U. Expert system support for the therapeutical management of cerebrovascular disease. Artificial Intelligence in Medicine, 2: 1990, 35-42. Müller-Kolck U. Diagnostic Consultations: A Speech-Act Theoretical Reconstruction fort he Design of Consultations Systems. Methods of Information, 1991, 311-315. Müller-Kolck U. Expertensysteme als metadiagnostische Hilfsmittel in ärztlichen Entscheidungsprozessen, in: Hucklenbroich P, Toellner R (eds.) Künstliche Intelligenz in der Medizin. Klinisch-methodologische Aspekte medizinischer Expertensysteme, Stuttgart, Jena, New York: Fischer Verlag, 1993, 141-159. Müller-Kolck U. Basic structures of nosology in medical arguments (in german), in: Meggle G., Nida-Rümelin J., Perspektiven der Analytischen Philosophie, Band 18, Proceedings of the 2nd Conference „Perspectives in Analytical Philosophy", Vol III, Berlin, New York: De Gruyter 1997, 518-528. Müller-Kolck U. Modellierung individueller Prognosen in der klinischen Medizin. Theory Biosci, 120: 2001, 45-56.
[5] Suppes P., Introduction to Logic. New York, Cincinnati, Toronto, London, Melbourne: Van Nostrand Reinhold Company, 1957.
[6] Balzer, W., Moulines C. U., Sneed, J.D.: An Architecture for Science, The Structuralistic Program, Dordrecht: D. Reidel Publ., 1987.

clinical medicine are reconstructed. The technique of defining set theoretical predicates.is self-explanatory, the formal representation uses naïve set theory.

This approach has three advantages: (1) The predicates reconstruct and summarize basic structures and their functions in clinical medicine. (2) They can be used to elicit basic heuristics of medical knowledge processing. (3) During the initial phase of a knowledge engineering process this kind of intuitive representation helps to axiomatize the medical domain in a precise manner without recourse to specialized formal languages which later on are necessary to transform the modelling into computer programs.

3 Patient Images

This chapter introduces the basic structures for reasoning in the practice of clinical medicine. To simplify matters some abstract terms like "patient self image", "partial potential patient image" or "potential patient image" are proposed. They are explicated on a very general level. The "image"-term stresses the "iconographic" view of the physician on his patient. From the viewpoint of cognitive science these patient images may be interpreted as the physician´s internal cognitive representations of individual patient data.

The iconographic view is predominant in clinical reasoning. The interaction between patient and physician is based on images of the patient. The physician produces multiple images to capture the patient's problem. The old and often used analogy between physicians and artists illustrates the intuitive meaning of the iconographic view. This analogy compares the view of the physician on his patient with the view of the painter on his model. The painter creates an individual image. He transforms his sensual cognitive impressions into forms and colours on the screen, the portrait.

The portrait of the model resembles the medical patient image whose features are documented by the physician in multiple clinical data records. Interactions between physicians and their patients are based on similar iconographic acts. Like the painter the physician portraits his model, i. e. his patient. The physician creates an internal cognitive image representation. These images are often classified as mosaics, because they are puzzled together from single pieces of medical data. The physician uses incomplete and fuzzy clinical data sets to create a picture of the ill subject in its individual domain. This approach is called the ***axiom of iconographic reasoning in clinical medicine***.

All interactions between a physician and his patient are influenced by their self-images. First, the patient has subjective intentions. He wants to achieve a specific result by consulting the doctor. The physician on the other side produces his personal individual image of the patient. This image depends on the doctor's own intentions, preferences, on his medical knowledge, and on his technical skills. In clinical medicine all interactions are deeply rooted in both the patient's self-image and in the doctor's self-image. This is the ***axiom of interacting self-images***.

The analysis of the basic structures of clinical reasoning starts with the patient's self-image, abbreviated with **PSI**, and the doctor's self-image, abbreviated with **DSI**.

3.1 Patient Self-Image (PSI)

Patients are often vague in their communications. Issues and facts of their problems are hidden by his / her personal interpretations of disease signs. Fears and worries disguise his true situation. Successful communication depends on the linguistic abilities of the patient. On the one hand, extremes of such communications are illustrated by a person, who exaggerates his problem in a never ending tale of woe and, on the other hand, by a disabled person, who is unable to speak or to act. In the first case, the physician has to listen to the patient and has to separate relevant from less relevant facts. In the second case he observes the patient for a longer time and consults medical attendants and nurses to get a correct image of the patient's problem. Thus, the physician is confronted with open and hidden contexts. He has to understand the individual life situation of the patient.

History taking is the basis of clinical medicine. In contrast to scientific investigations, which refer to the description of classes, historical reflections refer to individual events or individual cases. History taking leads to a first image of the patient. The history of the patient's disability focuses

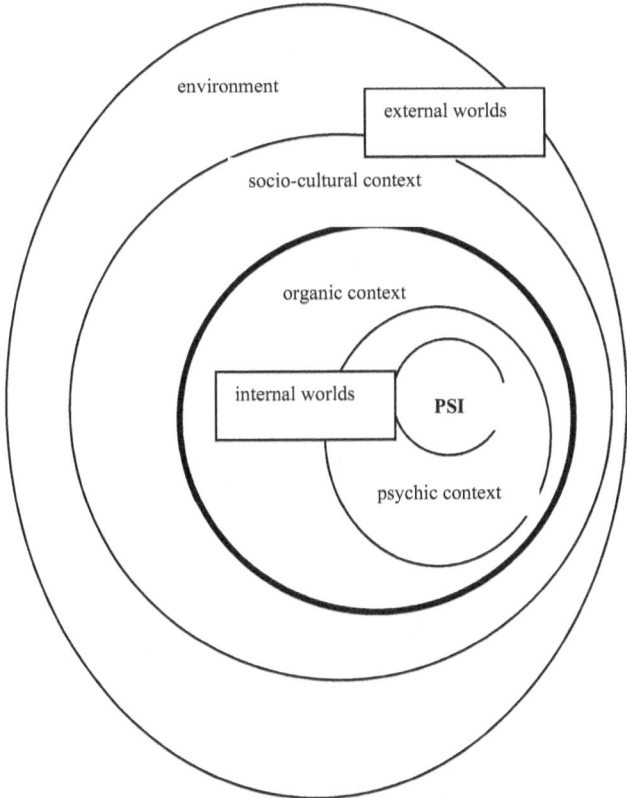

Figure 1: Four contexts for an individual patient self-image **PSI**. **PSI** contains everything that the patient believes or knows about himself and his internal and external worlds. Environment and socio-cultural context form the external world for **PSI**. Organic and psychic context form the internal world for **PSI**.

on the patient's account of his current illness, previous illnesses and his general state of health, family history, social history, and psychological assessment. Whereas organic and psychic contexts represent the "internal" world of a patient, the socio-cultural and environmental contexts constitute his "external" world (figure 1). The patient is part of environmental and socio-cultural contexts as well. In order to understand and interpret events of his internal and external worlds and their interactions each patient uses his individual personal knowledge. These individual interpretations connect the two domains of his internal and external world. Using this knowledge the patient creates an image of his own, a self-image for private use and for the doctor.

The physician perceives the patient's self-image by observing the patient and listening to the patient's spontaneous verbalisations. The patient self-image represents facts and interpretations which the patient expresses spontaneously or by being requested to do so. The patient utters his troubles and tells his interpretation why he feels sick. The patient uses a personal value system for his medical interpretation of his situation. He refers to personal values to classify his situation as normal or beyond his personal conception of normality (figure 2).

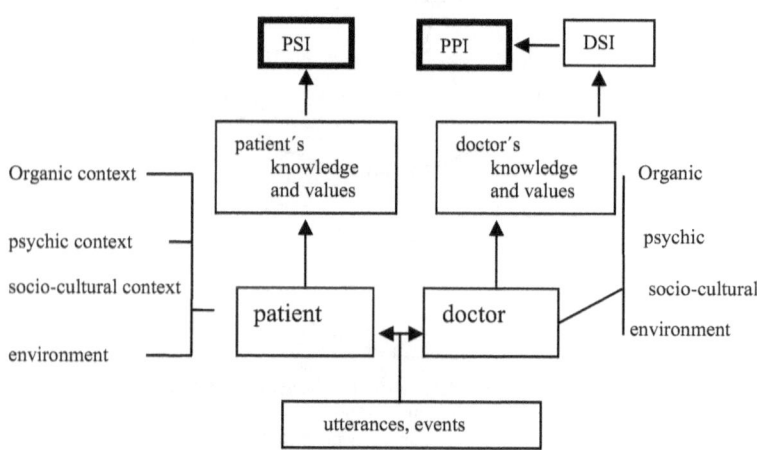

Figure 2: Contexts in the physician-patient-relation. Utterances and events represent the patient-doctor interactions. The patient self-image (**PSI**) represents the patient's interpretation of his situation, the potential patient image (**PPI**) represents the doctor's interpretations. In the concrete clinical context both patient and doctor belong to the same environment, they may share the same social-cultural context. DSI = doctor's selfimage.

The patient self-image represents a very specific and individual domain. It contains individual features of the person. The patient uses idiosyncratic verbal and nonverbal symbols to describe his personal view of his world, his special situation, his concrete health problem. His descriptions are influenced by the different organic, psychological and socio-cultural contexts to which he belongs. The patient interprets his situation as healthy, indifferent or sick by using his personal knowledge of the different contexts and based on his personal values. The basic structure of a patient self-image (**PSI**) I define as follows:

Definition 1: PSI is a set of patient self-images, iff for C_{org}, C_{psy}, C_{soc}, C_{env}, V_{pat}, NV_{pat} there are the functions v, nv, δ such that:
1. **PSI** = $\langle C_{org}, C_{psy}, C_{soc}, C_{env}, V_{pat}, NV_{pat}\ v, nv, \delta \rangle$,
2. C_{org} is a set of organic contexts,
3. C_{psy} is a set of psychological contexts,
4. C_{soc} is a set of socio-cultural contexts,
5. C_{env} is a set of environmental contests,
6. v: $C_{org} \times C_{psy} \times C_{soc} \times C_{env} \rightarrow V_{pat}$,
7. nv: $C_{org} \times C_{psy} \times C_{soc} \times C_{env} \rightarrow NV_{pat}$,
8. for all C_{org}, C_{psy}, C_{soc}, C_{env}:
   δ: $V_{pat} \times NV_{pat} \rightarrow$ {healthy, indifferent, sick}.

A patient describes his problem with verbal or non-verbal utterances. Both, the spontaneous verbalisations about his soundness and his non-verbal utterances create an initial patient-image. In the definition above verbalisations are represented as a function v with four domains of the different worlds or contexts C_{org}, C_{psy}, C_{soc} and C_{env}. The set V_{pat} is the co domain for the function v. V_{pat} contains the subjective views of the patient on his health problem. The set V_{pat} usually represents patient data, which can be found in the verbalized patient's history.

All verbalisations of a patient are accompanied by non-verbal utterances. They can be recognized by changes of the patient's behaviour or of his appearance. Non-verbal utterances can be observed even if the patient remains completely silent. In Definition 1 non-verbal utterances are defined as a function nv with the four domains C_{org}, C_{psy}, C_{soc}, C_{env}.

The functions v and nv are emergent functions of the human nervous system which is part of the constituting elements of C_{org} and C_{psy}. The socio-cultural context C_{soc} determines how the patient utters his feelings. His cultural imprints exease an influence the modality of his utterances. His linguistic socialization determines the modality of his verbal utterances. His general knowledge and other sorts of specialised knowledge are elements of C_{soc}. This general and specialised knowledge is necessary for the interpretation function δ. I simply regard the function δ as a given constant of human reasoning.

Although V_{pat} und NV_{pat} refer to the same domains, both sets are not identical. They are often very different. For instance, an obviously ill person who pants for air may assert that he is not diseased. Although NV_{pat} contains classical disease signs showing that the patient is panting for air V_{pat} contains verbalisations suggesting that the patient subjectively feels not ill. One motive for this discrepancy may be that the patient dissimulates his true situation, for example, because he fears to be sent to the hospital. Vice versa, a person who is completely well may feel critically ill and demands to be send to hospital, because he fears to suffer from a lethal disease although all objective data prove the contrary.

Based on the set of individual data sets V_{pat} and NV_{pat} the function δ produces an interpretation of the concrete situation of the patient. Function δ is the interpretation function which says whether the patient regards himself being ill or not. Sometimes this decision is uncertain, and the patient is not able to decide between both possibilities. Then he remains indifferent with regard to a concluding interpretation. The meaning of "ill" and "healthy" in **PSI** refers to the patient's personal interpretations and not to professional medical knowledge.

The patient uses verbal and non-verbal utterances to describe his actual life-situation. The patient produces a picture of his actual setting. Something very similar happens with the doctor.

3.2 Doctor Self-Image (DSI)

The physician has no direct access to the patient's world. The physician must interpret utterances of his patient. Subsequently he uses medical test to confirm his interpretations and assumptions about the patient's world. The physician starts to study the patient self-image **PSI**. His aim is to get a personal and complete reproduction of **PSI**.

In essence the physician seeks to get to know and to understand the signs and symptoms presented in the patient self-image **PSI**. In doing so he relies on his personal experiences, his general knowledge and his specialised professional medical knowledge. The doctor's world is schematically opposed to the patient's world in figure 2. All these features of the doctor's world define the doctor's self-image **DSI**. Obviously, the structure of **DSI** is very similar to that of **PSI**.

Definition 2: DSI is a set of doctor self-images iff for C_{org}, C_{psy}, C_{soc}, C_{env}, V_{doc}, NV_{doc} there are the functions v, nv, δ such that:
1. **DSI** = $\langle C_{org}, C_{psy}, C_{soc}, C_{env}, V_{doc}, NV_{doc}\ v, nv, \delta \rangle$,
2. C_{org} is a set of organic contexts,
3. C_{psy} is a set of psychological contexts,
4. C_{soc} is a set of socio-cultural contexts,
5. C_{env} is a set of environmental contests,
6. v: $C_{org} \times C_{psy} \times C_{soc} \times C_{env} \to V_{doc}$,
7. nv: $C_{org} \times C_{psy} \times C_{soc} \times C_{env} \to NV_{doc}$,
8. for all C_{org}, C_{psy}, C_{soc}, C_{env}:
   $\delta: V_{doc} \times NV_{doc} \to \{healthy, indifferent, sick\}$.

The biological, psychological, socio-cultural and environmental contexts of the patient and of the physician are structurally similar. If the doctor talks about his own sufferings, he is in the same situation as the patient. The index *doc* indicates that **V** and **NV** contain personal verbal or non verbal utterances of the doctor.

When the doctor tries to understand the complaints of his patient, the function δ produces interpretations of the health status of his patient. Within the same structure the physician uses V_{pat} and NV_{pat} of **PSI** to decide whether the patient's status is healthy, indifferent, or sick.

The contexts C_{org}, C_{psy}, C_{soc}, C_{env} can be similar for the patient and the physician, or they can be completely different. How much they overlap influences their mutual understanding and their actions. Either the physician immediately recognizes the problem of his patient and decodes the patient's symptoms and signs or all utterances of the patient are completely incomprehensible and seem to appear strange. Between both extremes all transitions of good or bad communication are possible.

The main difference between the contexts C_{org}, C_{psy}, C_{soc}, C_{env} of the patient and those of the physician refer to the professional medical knowledge of the physician, which is part of the socio-cultural context C_{soc}. Professional medical knowledge and technical medical skills are the characteristic parts of **DSI**.

The specific difference between **PSI** and **DSI** is revealed by a deeper look into the structure of medical knowledge. In the next chapter I analyse medical descriptions and basic elements of the nosology. This leads to generic set of nosology elements **NE** and their connection to nosology nets **NN**.

3.3 Medical Descriptions

Medical objects are most suitably represented in a decomposition hierarchy.[7] Table 1 below shows the consecutively numbered levels of such a hierarchy. J. Rasmussen has proposed five layers for the abstraction hierarchy.[8] They

---

[7] Blois MS. Information and Medicine, The Nature of Medical Descriptions, Berkeley: Univ. of California Press, 1984.
[8] Rasmussen J. Modelling distributed decision making, in: Rasmussen J, Brehmer B, Leplat J (eds.): Distributed Decision Making: Cognitive Models for Cooperative Work, John Wiley 1991, 111-142.

specify the functions of the objects assigned to the different levels of decomposition.

The medical decomposition hierarchy has ten levels. The first level D1 of the decomposition hierarchy is the holistic level. At this level the patient is regarded as a whole. Nosological descriptions of several patients or groups are possible as well as the description of social systems and of patients within their social relations. At the second level D2 larger interrelated anatomical parts of the human being are described, e.g. chest or abdomen. Physiological systems, e.g. cardiovascular system, alimentary system, genito-urinary system, nervous system, locomotor system belong to the third level D3. The parts of these systems, e.g. organs as heart or lung are described at the fourth level D4. At the fifth level D5, parts of the organs are described, e.g. heart valves, pulmonary lobe. Single cells and cell structures are described at the sixth level D6, parts of the cell at the seventh level.

	A1	A2	A3	A4	A5
D1 / level 1					
D2 / level 2					
D3 / level 3					
D4 / level 4					
D5 / level 5					
D6 / level 6					
D7 / level 7					
D8 / level 8					
D9 / level 9					
D 10 / level 10					

Table 1: AD-table: Abstraction hierarchy of A1 = form, A2 = function, A3 = macro-function, A4 = abstract function, and A5 = purposes; decomposition hierarchy D1 – D10 of ten levels of description of medical objects.

Level D8 and level D9 are denoted to macro- and micro-molecules. Level D10 refers to atoms and ions.

The abstraction hierarchy starts with the first layer A1 describing the biological form of the medical object. It is consistent with the corresponding level of the decomposition hierarchy.

The next layer A2 refers to the biological function of the objects. The functions are very different. This depends on their domains, i. e. the decomposition level of medical objects.

Macro-functions establish the third layer A3. Macro-functions show the interactions between objects of the second layer, i. e. blood pressure or digestion. Blood pressure is a function of the objects heart, vessels, blood, hormone level, autonomous nerves. Digestion is a function of the mouth, oesophagus, bowels, digestive enzymes, autonomous nervous function and so on.

On the fourth layer D4 we find abstract functions, e.g. feed-back regulation as applied in the description of blood pressure regulation, respiratory regulation or temperature regulation.

Deontic rules for medical decision making, intended purposes in the context of a means-ends analysis or medical norms for an individual risk-benefit analysis belong to the fifth layer A5.

The decomposition hierarchy and the abstraction hierarchy form the nosological object matrix for nosological nets **NN**, whose structure is described next.

3.4 Nosology Entity (NE) and Nosology Nets (NN)

Disease entities are the generic units of the medical nosology. In the following they are referred to as nosology entities **NE**. **NE** denotes an autonomous knowledge unit that can be used to formulate reasonable diagnostic and therapeutic propositions (figure 3). **NE** is defined as follows:

Definition 3: **NE** is a nosology entity, iff there are a nosology concept **NC** and a set of intended medical applications **IMA** such that
NE =⟨**NC, IMA**⟩.

Nosology concepts $\{nc_1,..., nc_n\}$ are abstract descriptions of a disease. They form the set **NC**. Intended medical applications are typical clinical examples of a disease which illustrate the range and meaning of a nosology concept nc_n. **NE** is an ordered pair of **NC** and **IMA**, **NE** = ⟨**NC, IMA**⟩. **NC** is necessary for the proper identification and classification of diseases. The elements in **NC** are results of scientific medical research. **NC** defines, what is pathologic and what is normal[9]. **NE** contains the law like rules and empirical generalisations represented in the systematic nosology or in pathophysiological theories. The diagnostician causally interprets functional relations with the help of this medical knowledge.[10]

IMA contains prototypical examples for **NC**. The logical structure of **NC** is a specific topic of the philosophy of science.[11]

---

[9] For the connotations of normality: Murphy E.A., The Logic of Medicine, Sec. Edition, Johns Hopkins Universitiy Press 1997, p143-155. For the meaning of inteded application: Balzer, W., Moulines C. U., Sneed, J.D., a.a.O. Sneed JD, The Logical Structure of Mathematical Physics, Dordrecht: Reidel Publ. Co., sec. edition 1979.
[10] Sadegh-Zadeh K, The notion of disease and nosological system (in german, Metamed 1, 1977, 4-41. Sadegh-Zadeh K. Foundations of clinical praxiology, part II: categorial and conjectural diagnosis, Metamedicine 1982: 159-191.Sadegh-Zadeh K. Fundamentals of clinical methodology: 1. Differential indications, Artificial Intelligence in Medicine 6, 1994, 83-102. Sadegh-Zadeh K. Fundamentals of clinical methodology: 2. Etiology, Artif. Intell Med 1998: 227-270.
[11] Hucklenbroich P. Steps towards a theory of medical practice, Theor Med. Bioeth. 19, 1998: 215-38. Müller-Kolck U. Basic structures of nosology in medical arguments (in german), in: Meggle G., Nida-Rümelin J., Perspektiven der Analytischen Philosophie, Band 18, Proceedings of the 2nd Conference „Perspectives in Analytical Philosophy", Vol III, Berlin, New York: De Gruyter 1997, 518-528. Murphy EA. Skepsis, Dogma and Belief. Uses and Abuses in Medicine. Baltimore, London: The Johns Hopkins University Press, 1981. Sadegh-Zadeh K, The notion of disease and nosological system (in german, Metamed 1, 1977, 4-41.

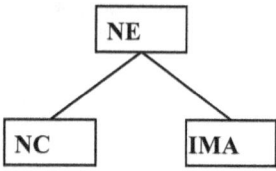

Figure 3: Structure of a nosology entity

Elements of **IMA** have to meet special conditions to qualify for an intended application of **NE**. They are defined below.

Definition 4: **IMA** are intended applications of **NE** iff there are BS_{NE}, M_{NE}, DM_{NE} so that:
1. $IMA = \ <BS_{NE}, M_{NE}, DM_{NE}>$;
2. BS_{NE} is a not empty basic set;
3. M_{NE} is a set of manifestations;
4. DM_{NE} is a set of diagnostic manifestations,
5. $DM_{NE} \subseteq M_{NE}$;
6. **IMA** performs the causal rules of NC_{NE}.

The basic set BS_{NE} contains medical objects from the first column of the medical decomposition- and abstraction hierarchy (see table 1).

Manifestations M_{NE} of a disease entity **NE** are signs or symptoms. Signs are primary phenomenon of the disease. The physician can directly perceive them without medical technology. These manifestations are called "clinical manifestations". The term "sign" must be distinguished from the term "symptom". Symptoms are subjective disorders as reported by the patient. If the patient feels sick, his condition is represented by the noun "sickness" or "illness". In contrast, the term "disease" is used to denote his objectively diagnosed nosological condition. The other manifestations of a disease can only be detected with the help of medical technology, e. g. examination of urine and blood, radiography.

Diagnostic manifestations DM_{NE} are a subset of manifestations M_{NE}. The elements in DM_{NE} are identifiers. They are so called "pathognomonic" features for **NE**. Pathognomonic signs are not always present. Only three of four diseases show pathognomonic signs.

Diagnostic manifestations are specific features of **NE**. They can either be detected by clinical examination or with the help of medical technology. At the end of the diagnostic process they help to detect their corresponding **NE**. In other words, the meaning of DM_{NE} is **NE**–relative.

DM_{NE} is defined as follows:

Definition 5: An attribute of an element of the basic set BS_{NE} is a diagnostic manifestation $dm \in DM_{NE}$ with reference to **NE**, iff each determination of dm in an application of **NE** presupposes at least one concrete intended medical application $ima \in IMA$ of **NE**.

The elements in DM_{NE} are not always distinct and clearly distinguishable from non-diagnostic manifestations. A characteristic feature of DM_{NE} is its fuzziness. As a consequence the difficulties of differential diagnosis are due to this fuzziness.

Nosology entities **NE** are interconnected by a specialisation relation S. It connects several sets of **NE** to a nosology net **NN** (figure 4).

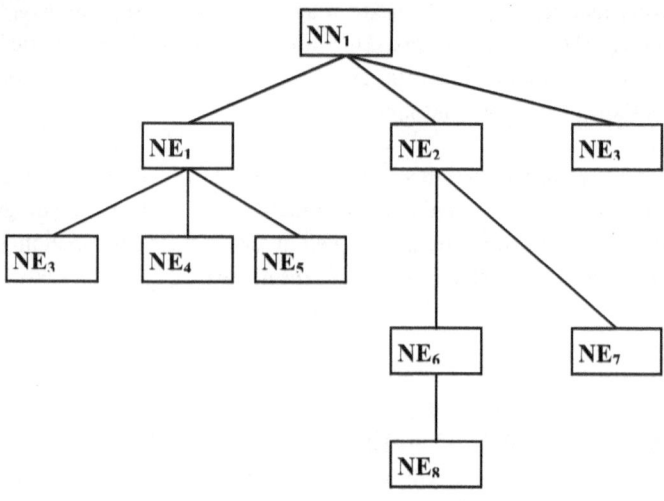

Figure 4: Example of a nosology net

A nosology net **NN** is defined as follows:

Definition 6: **NN** is a nosology net iff
1.   **NN** = < Pot (**NE**), S>;
2.   Pot (**NE**) is a non empty power set of **NE**;
3. $S \subseteq$ Pot (**NE**) × Pot (**NE**) is a specialisation.

The decomposition hierarchy of the basic set BS_{NE} determines the number levels of medical descriptions in **NN**.
After clarifying the specific difference between **PSI** and **DSI** by looking into the structure of medical knowledge the structure of patient images is analyzed in the next chapter. It is argued that the interaction between **PSI** and **DSI** leads to different patient images which the physician uses for his clinical work.

3.5 Patient Image (PI)

An exact replica that is indistinguishable form the original **PSI** is called patient image **PI**. **PI** is to be understood as a copy of **PSI**. Then the structure of the patient image **PI** is identical with **PSI**. **PI** is a complete mapping of **PSI**.

All the epistemological constraints which obviate that the physician acquires **PI**, a complete understanding of his patient, are not discussed here. It is self-evident that the ideal of a complete understanding of the patient cannot be achieved by the diagnostic process.

Coming back to the artist-physician analogy it is well known that portraits qualitatively differ compared with the true model. Their granularity differs. They never catch the true person. These epistemic limitations also apply to clinical medicine. To capture the "true" image of the patient remains a fiction. For the ongoing analysis it is simply presupposed that **PSI** = **PI** is impossible.

Images of a patient in the sense of so called "hard" and "soft" medical data are always approximations. This limitation is called the ***approximation axiom***. This axiom leads to the differentiation between partial potential patient images, abbreviated with **PPPI**, and potential patient images, abbreviated with **PPI**. **PPPI** is described at first.

3.6 Partial Potential Patient Image (PPPI)

During the initial contact the physician gains partial insight in the patient's world and therefore he only receives a partial potential reproduction of the patient self-image **PSI**. This partial copy of **PSI** is called partial potential patient image **PPPI**, which is defined as follows:

Definition 7: **PPPI** is a set of partial potential patient images, iff for **DSI**, V_{pat}, NV_{pat}, **PL** there are the functions σ and π such that:

1. **PPPI** = < **DSI**, V_{pat}, NV_{pat}, PL_{pat}, IP_{pat}, σ, π>,
2. **DSI** is a set of doctor self-images,
3. σ: $DSI \times V_{pat} \times NV_{pat} \to PL_{pat}$,
4. π: $DSI \times PL_{pat} \to IP_{pat}$.

The initial history taking ends with a list of patient problems the physician has to solve.[12] The problem list forms a set denoted as PL_{pat}. The index indicates that **PL** refers to an individual patient.

The function σ is a PL-generation function that produces PL_{pat} from the available non-verbal and verbal utterances of the patient in the data sets V_{pat} und NV_{pat} and from the doctor's professional and general knowledge represented in **DSI**.

Five constraints apply to the PL-generation function σ:

1. *Specificity constraint*: Describe a problem in **PL** as precise as possible in common speech, so that it can be reappraised by the patient.
2. *Sensitivity constraint*: Avoid meticulous description.
3. *Completeness constraint*: Try to get an extensive **PL**.
4. *Interaction constraint*: Describe possible interactions of the problems in **PL**.
5. *Acuteness constraint*: Decide on the acuteness of problems.

Based on **PL** the doctor generates initial plans, which are denoted by the set **IP**. Initials plans show the way how to solve the problems of the patient listed in **PL**. **DSI** and **PL** are the domains of the IP-generation function π. The doctor uses his professional medical knowledge, represented in C_{soc} of **DSI**, and the list of patient problmes **PL** for the generation of his initial plans. Initial plans are listed in the set **IP**. Usually **IP** contains diagnostic as

---

[12] Weed, L.L.: Medical Records, Medical Education and Patient Care – The Problem-oriented Record as a Basic Tool. Cleveland, Ohio: Press of Case, Western Revers University, 1970.
Weed, L.L.: Knowledge Coupling; New Premises and New Tools for Medical Care and Education, Berlin et al.: Springer-Verlag, 1991.

well as therapeutic plans. Diagnostic plans aim at the nosological classification of the patient's problem. Therapeutic plans support the management of the patient's problem. Some situations need immediate help prior to knowing the exact diagnosis. Other situations may be less threatening. Then, there is time enough to wait for the correct diagnosis and to start treatment. Initial plans are algorithmic routine jobs. They are basic elements of the partial potential patient image. Details of **IP** are diskussed in the chapter on "Heuristic Processing" below.

The diagnostic and therapeutic process transforms the partial potential patient image **PPPI** into the potential patient image **PPI**. The main difference between both is the successful nosological classification of **PPPI**.

3.7 Potential Patient Image (PPI)

The set of potential patient images **PPI** contains all the elements of **PPPI** in conjunction with a nosological classification of the patient's problems. The difference between **PPPI** and **PPI** can be compared with the difference between the presumptive diagnosis and the established diagnosis. In common clinical language use **PPPI** resembles the presumptive diagnosis, whereas **PPI** refers to an established diagnosis.

Definition 8: **PPI** is a set of potential patient images, iff for **PPPI** and **NN** there is the function θ such that:
1. **PPI** = < **PPPI, NN,** θ >,
2. θ: **NN** × **PPPI** → **PPI**,
3. **PPI** $\subseteq$ **IMA**$_{NE}$.

The central element of **PPI** is the nosological classification function θ. Based on the professional medical knowledge in **DSI** and on the initial patient data the function θ classifies the patient's problems into nosological categories. The domains of the classification function θ are **PPPI** and **NN**, its co domain **PPI**. **NN** is the set of nosology nets, which is represented in **DSI**. For short, the above definition only mentions the relevant element of **DSI**, i. e. **NN** representing the medical nosology in **DSI**. **NN** also contains the clinical and diagnostic manifestations of the patient after a sequence of clinical and technical examinations. The modus of clinical examination and the selection of tests are defined in **NE** by the specifications of **IMA**.

PPI reconstructs the very general structure of the clinical diagnosis. Whereas **PPPI** represents a broad set of possible patient images, **PPI** is the smaller set of possible classified images which are closer to the real disease of the patient.

If the set **PPI** is not empty, **PPI** contains one or more elements. If there is more than one element, **PPI** contains possible differential diagnoses. As **PPI** is not a static set, but changes during the diagnostic process, the physician has to consider these differential diagnoses in order to solve the patient's problems. If the patient suffers from more than one disease, then **PPI** should contain a list of these diseases at the end of the diagnostic process. If the patient suffers only from one disease, then **PPI** contains the list of differential diagnoses at a particular time of the diagnostic process.

The dimensionality of **PPI** depends on the number of involved nosology entities, i. e. the number of elements in **NN**. **PPI** can be one-dimensional or multi-dimensional. **PPI** is one-dimensional, if **NN** only contains one **NE**. **PPI** is multi-dimensional, if **NN** contains two or more, $NE_1 \times \ldots \times NE_n$. In this case, the medical events in **PPI** belong to different nosologies. **PPI** with $NN \subseteq NE_1 \times \ldots \times NE_n$ represents the clinical concept of multi-morbidity. **PPPI** is not always completely **NE**-classifiable (figure 5). In extreme cases all medical events in **PPPI** are not **NE**-classifiable. In this case, **PPI** is empty. **PPPI** is then partitioned into subsystems whose behaviour can be interpreted with the help of general principles from physiology or biochemistry. These theories are used to understand the disorders in **PPPI** tentatively. **NN** is substituted by general systems theories of pathophysiology, e. g. theories of heart dysfunctions or circulation dysfunctions or theories of disturbances in metabolism.

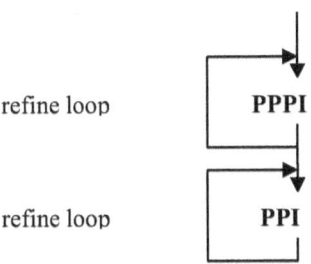

Figure 5: Movement of reasoning from utterances to **PPPI** or **PPI** with refine loops for **PPPI** and **PPI**. The refine loops improve the granularities of **PPPI** and **PPI**.

PPI is time dependent. Although its structure does not change, its cardinality changes over time.

Dynamics and kinetics of **PPI** are described by courses. The disease course is a basic source for medical data and a very important element of medical reasoning.

4 Disease Course

PPI is not a static information object. **PPI** represents "snapshots" of complex disease processes. Besides the diagnostic classification the central topic in practical medicine is the understanding of the individual kinematics and dynamics of **PPI**. Medical everyday language characterizes the variations of **PPI** over time as „disease course". The disease course is the essential core of clinical reasoning.

The course indicates an individual development of **PPI** over time, one of the most important sources of information for the physician.

Definition 9: **PPI-C** is a course of **PPI** iff there are **PPI**, **I** and < such that
1. **PPI-C** = ⟨**PPI**, <, **I**⟩,
2. **I** is a non-empty set of time intervals,
3. < is an antesessor relation in **PPI**.

If **PPI** is empty, **PPI** in the definition above is substituted by **PPPI**, until **PPPI** is **NE**-classifiable:

Definition 10: **PPPI-C** is a course of **PPPI** iff there are **PPI**, **PPPI**, **I** and < such that
1. **PPI** is empty
2. **PPPI-C** = ⟨**PPPI**, <, **I**⟩,
3. **I** is a non-empty set of time intervals,
4. < is an antesessor relation in **PPPI**.

A formal representation of time is a period structure ⟨I, ⊆, <⟩ with a partial ordering ⟨I, ⊆ ⟩, whose subset is a strict partial ordering < such that the triplet ⟨I, ⊆, <⟩ satisfy the condition on monotony. If there are two intervals i_1 and i_2 with i_1 and $i_2 \in I$, then $i_1 < i_2$ means „i_1 is earlier than i_2", or „i_2 is later than i_1".

The antecessor relation $<$ presupposes that time isomorphically to a straight line spans from the past to the future.[13]

The following explanations of the two definitions above refer to **PPI-C** and **PPPI-C** as well. Medical events in **PPI-C** as well as in **PPPI-C** are temporally ordered by the relation $<$ in the same way. In addition the antesessor relation is causally interpreted with the help of the medical nosology. Considering two events e_1 and e_2 then $e_1 < e_2$ means "e_1 causes e_2" or "e_2 is caused by e_1".

The function θ classifies individual patient data. As a consequence **PPI-C** is contingent and idiographic.

The antesessor relation in the definition of **PPI-C** is interpreted as an ordering of the potential patient images according to their linear chronology. **PPI** $= \{ppi_{t1} < ppi_{t2} < ppi_{t3} < ... < ppi_{tm}\}$ with $t_1, ..., t_n$ for the different time points. The elements in **PPI** are arranged in a linear structure with a chronological order.[14]

The set $\{ppi_{t1} < ppi_{t2} < ppi_{t3} < ... < ppi_{tm}\}$ represents a standard course of **PPI-C**. If the medical examination of the patient starts at ppi_{t3}, then $\{ppi_{t1} < ppi_{t2}\}$ represents the history of **PPI-C**. The utterances concerning the patient's history are interpretations of original events $\{e_1 < ... < e_n\}$ in the past. On the other hand the set $\{ppi_{t4} < ... < ppi_{tm}\}$ represents the future development of **PPI-C**. Usually three time intervals define the kinetics of **PPI-C**:

- the latency period of **PPI** (**PPI-C**$_{latency}$),
- the onset period of **PPI** (**PPI-C**$_{onset}$), and
- the duration of **PPI** (**PPI-C**$_{duration}$).

---

[13] Benthem, JFAK van. The Logic of Time, A Model-Theoretic Investigation into the Varieties of Temporal Ontology and Temporal Discourse, Dordrecht: Reidel Publ., 2nd Edition, 1992.
[14] Benthem JFAK van. A.a.O.

Thus, the inner time structure of **PPI-C** is **PPI-C**$_{latency}$ < **PPI-C**$_{onset}$ < **PPI-C**$_{duration}$.

The latency denotes the time from the very beginning of the disease process up to the detection of its first clinical manifestations M_{NE}. Usually the latency interval can only retrospectively be reconstructed from the known natural course of **PPI-C**, which is described in **NE**.

The onset of **PPI** follows the latency period. This is the time period from the first detection of elements in M_{NE} up to the time point where the correct diagnosis is establishes, i. e. there are sufficient diagnostic manifestations DM_{NE} to define **PPI**.

The duration of **PPI** is the third element of the inner time structure of **PPI-C**. This period, the duration of **PPI**, is usually characterized by the different disease states. States of **PPI-C** represent the severity of disturbances of **PPI** according to the natural course represented in **NE**.

Diseases represented in **PPI-C** can continuously progress, or they can reach partial or complete recovery ("restitutio ad integrum"). These features of their kinetics are represented by the natural course (**PPI-C**$_{natural}$). The natural course of a disease is the disease course without medical interventions. **PPI-C**$_{natural}$ is the essential starting point of each medical forecast. A prognosis is not reliable if **PPI-C**$_{natural}$ is unknown. Usually, **PPI-C**$_{natural}$ is divided into self-limiting and non self-limiting courses. Furthermore, **PPI-C**$_{natural}$ can be reversible, spontaneously irreversible and irreversible. **PPI-C**$_{natural}$ is spontaneously irreversible, if it is only reversible with external help. **PPI-C**$_{natural}$ is irreversible, if external interventions cannot completely or partially reverse the disease process. Some general constraints apply to **PPI-C**$_{natural}$:

- If **PPI-C**$_{natural}$ is reversible, then **PPI-C**$_{natural}$ is self-limiting.
- If **PPI-C**$_{natural}$ is spontaneously irreversible or permanently irreversible, then **PPI-C**$_{natural}$ is not self-limiting.

In every day medical practice the vague terms "acute", "subacute", and "chronic" approximately classify the dynamics of **PPI-C**. The term "acute" denotes a rapid and/or fulminate onset. The term "chronic" denotes a long-term condition or long-term course. The difference may be illustrated by two virus infections, influenza and hepatitis C virus infection. Influenza starts fulminantly and leads within hours to serious illness. Its course is fast and after about three weeks restitution is completed. Hepatitis C virus (HCV) infection, with over 170 million infected people worldwide the leading indication for liver transplantation in the United States and Europe, develops slowly.[15] Often patients do not know when they have been infected. A frequent extrahepatic complication of HCV is the HCV-induced mixed cryoglobulinemia.

The onset of this complication usually presupposes a time course of more than 20 years. Thus, disease dynamics are primarily very mild. The term "subacute" is defined for intermediate states between acute and chronic.

How **PPI-C** progresses, whether fast or slow, depends on two factors: the natural courses of **PPI**, i. e. **PPI-C**$_{natural}$, and on possible medical interventions.

For both, patient and doctor, it is the most interesting question how the disease develops in the future. Often, the patient is even more interested in the prognosis than in the correct diagnosis. Although he wants to know why he is ill he is primarily interested to know when he gets rid of his symptoms.

The future development of **PPI-C**, for short **PPI-C**$_{future}$, alludes to the complex topics of medical prognosis. Rationally justified medical forecasts are based on empirical generalisations, i. e. principles or rules which are represented in **NE**. By definition they are implemented in **PPI**. The expected future course is a conditional prognosis. It is a projection of the history of **PPI-C** into the future. The diagnostician presupposes that the future events in **PPI-C** are structurally similar to the past **PPI-C**.

---

[15] Charles ED, Dustin LB, Hepatitis C virus-induced cryoglobulinemia, Kidney international 76: 2009, 818-824.

In general medical prognoses are based on a ***homomorphism axiom of prognosis***: The expectation structure of the future **PPI-C** is homomorphic to the event structure of the past **PPI-C**.

For practical reasons of disease management the kinetics of future **PPI-C** is subclassified in short-term, middle-term, and long-term prognosis. If i_1 is the actual time interval and the prognosis refers to the next future time interval i_2, then the prognosis is called a short-term prognosis. This feature of a prognosis is represented by four relations (figure 6):

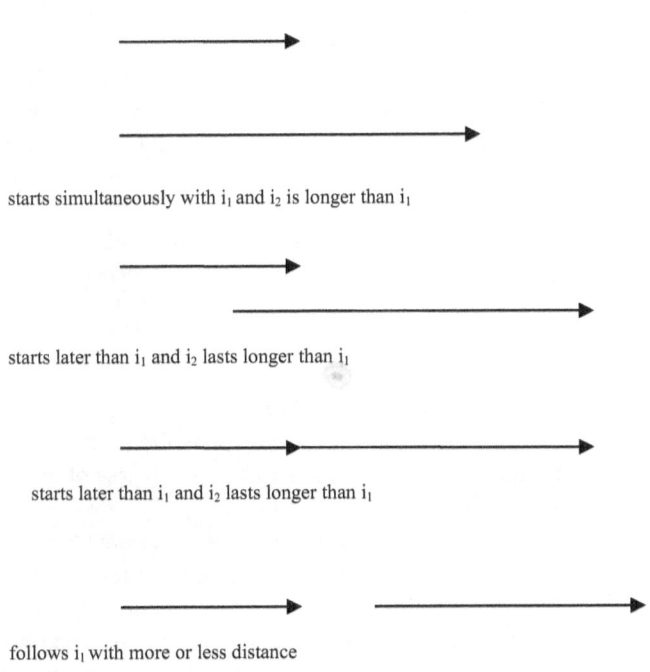

starts simultaneously with i_1 and i_2 is longer than i_1

starts later than i_1 and i_2 lasts longer than i_1

starts later than i_1 and i_2 lasts longer than i_1

follows i_1 with more or less distance

Figure 6: Four time relations for a prognosis

If the latency, onset and duration of the future **PPI** are known from **PPI-C**, the earliest and latest start of i_2, the earliest or latest end of i_2 and the earliest or latest end of i_1 can be estimated. If more than one future time intervals i_2, i_3, ..., i_n are regarded, more relations are necessary to represent a middle-term or long-term prognosis.[16]

The kinetics and dynamics of **PPI-C** depend on its dimensionality. The expectation structure of the prognosis is one-dimensional, if **PPI** is one-dimensional. In this case **PPI** is solely based on one nosology entity, i. e. virus hepatitis C. A relatively exact prognosis can be inferenced from **NE**. Variations of **PPI-C** are only due to individual modifications by the contexts in **PPI**.

A prognosis is multi-dimensional if it is deduced from a multi-dimensional **PPI**. In this case the prognostician has to decide, which **NE** out of a set $\{NE_1, ..., NE_n\}$ of relevant nosologies is primarily constitutive for **PPI-C**. If $\{NE_1, ..., NE_5\}$ with NE_1 = arterial hypertension, NE_2 = diabetes mellitus, NE_3 = hypercholesterinemia, NE_4 = chronic hepatitis C, NE_5 = coronary heart disease (CHD) are the relevant nosologies for **PPI**, then actual severity of CHD may be the most relevant factor for the future course of **PPI-C**. However, **PPI-C** is modified by co-morbidities as diabetes mellitus, hypercholesterinemia and hypertension. So the prognostician has to take into account the individual course of these entities for the individual prognosis of **PPI-C**.

Difficulties for prognostic reasoning arise if **PPI-C** turns out to be multi-dimensional. There is no empirically founded theory, how the interaction of different **NE** influences the individual course of **PPI-C** in the case of multi-morbidity. Forecasts are based on the individual **PPI-C**. Applying subjective probabilities in these cases presupposes examples from a reference class. Such a reference class is influenced by the selection bias of the prognostician. Predictions in the case of multi-morbidity are therefore usually represented as prognostic trends or scenarios. Prognostic trends or scenarios supercede exact prognostic inferences from a single **NE**. Predictions convert from rational **NE**-based prognoses to estimates about future trends of indi-

---

[16] Allen JF. Towards a general theory of action and time, Artificial Intelligence, 1984, 23: 123-154. Allen JF, Hayes PJ. Moments and points in an interval-based temporal logic, Computational Intelligence, 1989; 5: 225-238

vidual clinical situations. Predictions about trends depend on boundary conditions, which have to be clarified before inferencing a rational expectation structure for **PPI-C**.

The analysis of the patient-doctor interaction started with the simple model-painter analogy. Every day medical language uses terms like "disease pattern" or "clinical pictures". The analogies are model = patient, painter = doctor, portrait = patient image. Both the patient-self-image **PSI** and the doctor-self-image **DSI** determine the features of the patient image **PI**. The true **PI** is principally unknown. Good approximations for clinical work are **PPPI**. Medical knowledge is incorporated in **DSI**. Adjusted with medical knowledge **PPI** represents the diagnosed patient.

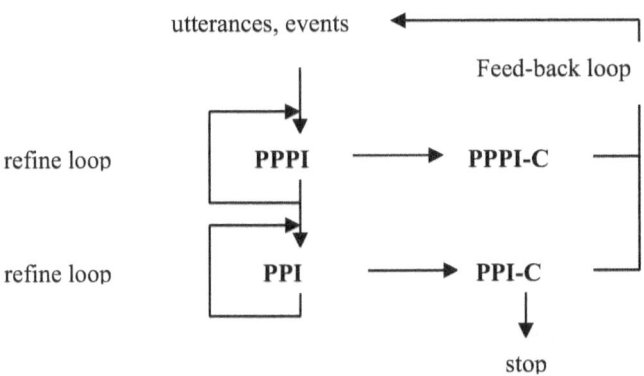

Figure 7: Movement of reasoning from utterances and events to **PPPI-C** or **PPI-C** and feed-back loop after changes of **PPPI** or **PPI**. If **PPPI-C** / **PPI-C** are reversible, movement stops. If **PPPI-C** / **PPI-C** are irreversible, movement continues until prognosis of PPI stops.

In the next chapter the heuristic processing of these conceptual objects is considered.

5 Heuristic Processing

The heuristics for processing the objects introduced above are now composed into three algorithms. The algorithms describe three basic cognitive tasks of the practice of medicine: history-based learning, diagnostic inferencing, and prognostic planning.

History-based learning is the initial task of the patient-physician interaction. It can be reconstructed as cross domain learning by task embedding. Its result is **PPPI**. Diagnostic reasoning aims at the production of **PPI**. Both **PPPI** and **PPI** incorporate historical aspects of the relevant medical events. In contrast to the historical view prognostic planning considers the future development of **PPI** or **PPPI**.

5.1 History Based-Learning

Each new patient confronts the physician with a new individual domain. To become acquainted with his new patient the doctor has to learn different things about the patient's world.

The physician initially sees each new patient as a new medical domain. The suitable method to learn new facts about a new patient is cross domain inference by task-embedding.[17] Cross domain inference is used to deal with new tasks by using solutions of already solved problems. The labyrinth example explains the method.

A successor follows a leader through a labyrinth which is unknown to the successor. The leader signs his way by symbols whose meaning is unknown to the successor. The successor however knows that he is solving a labyrinth problem. The successor must find the leader by the right interpretation of the signposts fixed at branchings by the ahead going leader. After some runs the successor has learned the meaning of two signs, i. e. a black square means

---

[17] Cummins R. Representation, Targets, and Attitudes, Cambridge Mass: MIT Press, 1996.

"follow me" or a black circle means "don't follow me". After further runs the successor uses his success in the first runs to expand his knowledge about the labyrinth and the signposts. For example, he learns the meaning of more complex signs as "the next branching turn right", "the next branching turn left". His problem solving strategy is embedded in the labyrinth task. It is the context for the search task.

When the doctor meets a new patient for the first time, his goal is to produce an initial **PPPI** of the patient's situation. **PPPI** is the goal of the doctor's learning task. Comparable to the labyrinth example of task-embedding the doctor uses knowledge about an already known domain **PPPI**$_1$ for the successful data acquisition of another domain **PPPI**$_2$. In other words, a task that demands knowledge about a patient world **PPPI**$_2$ is embedded in another task, which demands knowledge about a patient world **PPPI**$_1$ such that the success of the **PPPI**$_2$-task depends on the success of the **PPPI**$_1$-task.

Applying the labyrinth example to the medical settings the doctor can be compared with the successor who follows the patient into his personal idiosyncratic world to learn the symptoms of his illness and their meaning for the patient.

The doctor has to solve two tasks, (1) the acquisition of new knowledge about an unknown patient world and (2) the integration of the new **PPPI** into his nosological knowledge represented in **NN**.

In analogy to the labyrinth example the patient acts as the leader in the labyrinth whereas the physician plays the role of the searcher, who closely watches the new patient to understand and interpret his utterances. The utterances are comparable to the signposts from the labyrinth example whose meanings the successor tries to learn. Translated into common medical language the physician has to learn the meaning of the symptoms presented by the new patient. Even if the patient intends to communicate his problem successfully, at the beginning of the learning process typically both neither the patient nor the physician precisely know the real problem that leads to the consultation. Learning algorithms are usually based on two components: acquisition of new data and problem solving by integrating the new data into existing systems of knowledge. The following learning algorithm for the clinical context performs these components of knowledge processing. I call him **PPPI**-learning algorithm, because he describes operationally how the physician becomes acquainted with his new patient, how he learns a **PPPI**$_{new}$.

PPPI-learning-algorithm:
1. start patient-physician communication;
2. list the non-verbal **NV**$_{pat}$
3. list verbal utterances **V**$_{pat}$;
4. for **NV**$_{pat}$ and **V**$_{pat}$ find the corresponding patient contexts, **C**$_{org}$, **C**$_{psy}$, **C**$_{soc}$, **C**$_{env}$;
5. match $\langle$**C**$_{org}$, **C**$_{psy}$, **C**$_{soc}$, **C**$_{env}$, **V**$_{pat}$, **NV**$_{pat}$$\rangle$ with the understanding and intention of the patient;
6. σ: **DSI** × **V**$_{pat}$ × **NV**$_{pat}$ → **PL**$_{pat}$;
7. π: **DSI** × **PL**$_{pat}$ → **IP**$_{pat}$;
8. produce **PPPI**$_{new}$;
9. elaborate the history of **PPPI**$_{new}$;
10. match **PPPI**$_{new}$ with **PPPI**$_{old}$;
11. iterate algorithm until **PPPI**$_{new}$ is unchanged over
12. defined time period $[t_1, ..., t_n]$.

Explanation of the algorithm: The first steps of the algorithm are self-explanatory. Step 4 and step 5 need some annotations. A successful learning procedure presupposes a close cooperation between patient and physician. Cooperative communication is essential. As the labyrinth example shows the leader has to keep in mind that the successor has to learn from the leader. Sometimes it might be necessary that the leader assists the successor. The normal patient-physician interaction is comparable to this feature of the labyrinth setting. It is easier to understand the patient's problem if the patient cooperates with the physician. Often, however, the patient is not aware of this responsibility to the full extent. Therefore the physician has to make sure that his understanding of the patient's utterances really corresponds to the patient's own understanding of the same utterances. If the physician wants to constitute an undisturbed patient-physician-relation he should spend some time on this point before stepping ahead with the algorithm.

Step 6 and 7 are necessary by the definition of **PPPI**. Step 6 leads to the problem list **PL**. Based on the problem list step 7 produces initial plans **IP**.

IP contains two sorts of initial plans. The first sort of initial plans deals with the refinement of already collected data for **PPPI**. These plans are necessary to iterate the **PPPI**-algorithm until **PPPI** achieves a certain stability. The heuristic rule for the plan generation is the *Search-for-disease-indicators*-rule. Indicators are more reliable to detect an underlying disease. Parallel to the refinement of **PPPI**$_{new}$ diagnostic plans are generated. They are necessary to complete data collection to that extend which is necessary to start the diagnostic reasoning heuristic. Both, step 6 and 7, lead to step 8 that is the production of **PPPI**. It is important to emphasize that the set **PPPI** represents images of an individual patient, not images of a class of patients or a population.

Step 9 covers the important point of history taking in the practice of medicine. This step represents the historical view on the new patient. In contrast to the scientific inquiry which aims to describe classes of events, historical considerations in the practice of clinical medicine focus on individual events and their changes over time (figure 8). The arrows in figure 8 symbolize the kinetics and dynamics of **PPPI** during the past. The considered medical events of the example in figure 8 are blood pressure, blood sugar, blood fat, and pain of different parts of the patient's body, chest pain, pain of the legs, pain of the bones. These events are elements of the history of the patient. Figure 8 below also indicates that **PPPI-C** is simultaneously combined with aspects of the future development of **PPPI-C**. However, at the time point of the first evaluation of the patient the history is more important. Future aspects of **PPPI-C** are only necessary, if **PPPI** turns out to be not **NE**-classifiable (see step 15 of the diagnostic-inferencing-algorithm).

In figure 8 the y-axis is omitted. On the y-axis we see the elements of **PPPI-C**. The different disease stages are not explicitly mentioned. They are indirectly indicated by the time arrows. The slopes of the time arrows indicate the dynamics of the disturbances. The arrows in figure 8 have to be interpreted as follows:

- *blood pressure* shows a completely reversible course after one exacerbation;
- *blood sugar* progresses in two steps irreversibly;
- *pain in leg* is reversible two-times;

- *bone pain* progresses in two stepwise exacerbations as well as *chest pain*.

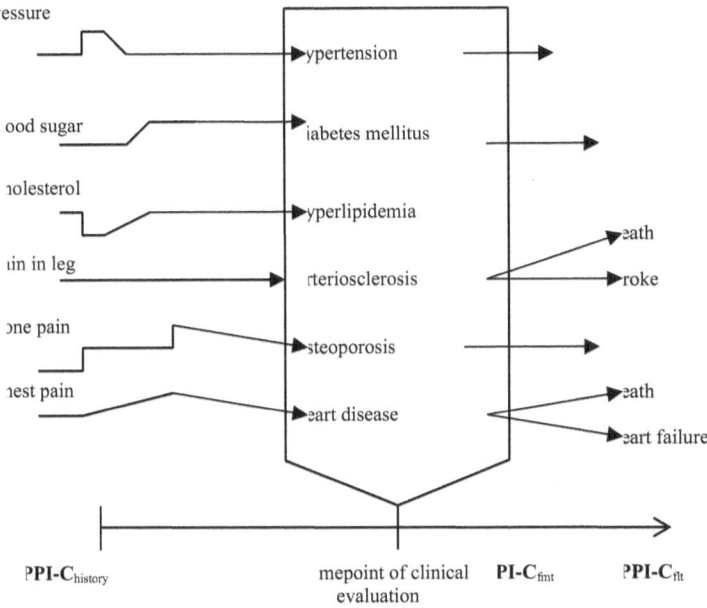

Figure 8: Time course of **PPPI-C**. fmt= future middle-term, flt = future long term. The arrows symbolise the kinetics and dynamics of **PPPI-C** for the medical events *blood pressure, blood sugar, blood fat, pain in leg, bone pain, chest pain*.

Step 10 of the algorithm summarizes the task-embedding procedure. The fragments of **PPPI**$_{new}$ are put in the context of already represented histories and facts about other patients. This step has to answer the question: "What can I already know about the new patient in comparison to other patients I have treated before?"

Step 11 deals with the non-historical aspects of work time constraints. Actions are time dependent. Pragmatic reasons presuppose that after some time of investigation **PPPI**$_{new}$ is regarded as stabile. It does not change during a defined suitable period of time. These pragmatically defined time periods depend on the natural courses of the medical processes and on the time management of diagnostic processes.

The following aspects have to be considered as inherent features of **PPPI**$_{new}$. The data in **PPPI** are incomplete with regard to **PI**. Above it is stated that a true **PI** is epistemologically impossible. The reliability of the represented signs and symptoms is unknown. The clinical data are uncertain. A basic axiom of the diagnostic reasoning is the incompleteness of data represented in **PPPI** at the end of the **PPPI**-learning algorithm and before starting the diagnostic inferencing.

Before starting diagnostic inferencing **IP** and **PL** generate diagnostic activities (clinical and technical examinations of the patient) which are defined in **NE** by the specifications of **DM**$_{NE}$.

5.2 Diagnostic Inferencing

The basic goal of diagnostic reasoning is the nosological classification of the individual **PPPI** and the approximately correct estimation of the inherent uncertainty of **PPI**. Function θ performs the nosological classification. Several preparatory steps form the diagnostic process. The basic principles are summarized in the diagnostic-inferencing algorithm.

Diagnostic-inferencing-algorithm:

1. nil nocere;
2. start with **PPPI**;
3. select diagnostic tests according to their power to detect strongdisease indicators,
4. consider existing diagnostic guidelines;
5. reference the specificity and sensitivity of the chosen diagnostic tests;
6. perform the tests sequentially or parallel depending on the acuity of the problem;
7. reference the disease base rates detected by the corresponding tests, if unknown, estimate them;
8. evaluate the pre-test and post-test probabilities of test results;
9. perform nosological classification function θ;
10. make a diagnostic decision and define **PPI**,
11. if step 9 fails, iterate algorithm until **PPI** is believable;
12. specify the dimensionality of **PPI**;
13. start prognostic planning algorithm;
14. refine **PPI** as long as its prognosis improves;
15. defend **PPI** against differential diagnosis,
16. if step 15 fails, iterate algorithm until **PPI** can be defended,
17. if step 16 fails, watch **PPPI-C**.

Explanation of the algorithm: Step 1 reconsiders the general principle that all medical actions should be primarily harmless for the patient. The expression "Nil nocere" can be translated with "primarily do not harm the patient". It is the Latin version of the ancient Hippocratic imperative to minimize the risk-benefit relation of all medical measures for a patient.

Step 3 refers to the selection of the diagnostic test. Step 3 presupposes knowledge of diagnostic test theory, which defines reliability and validity of tests. Their correct interpretation bases on additional information. **PPPI** contains information concerning the selection of diagnostic tests. **PPPI** provides the pre-test or a-priori probability for the diagnostic tests. The pre-test probability corresponds to the base rate of the detected disease in the patient population. Age-dependent disease base rates which refer to the actual population the physician usually has to care for are optimal. Unfortunately, these base rate data are often unknown. **PPPI** provides the baseline probability of the estimated disease. Quantitative tests are dichotomized into "positive" or "negative" by a discrimination value or cut-off-level which must be known by the diagnostician. Although tests are selected according to their power to detect the strongest disease indicator, the test procedures have to respect step 1 of the algorithm. According to this principle the pre-test probability of an invasive test, which principally can harm the patient, has to be enhanced by previous non-invasive test.

Step 4 and step 5 emphasize that existing guidelines for the selection of diagnostic tests have to be consulted before ordering the test as well as the sensitivity and specifity of the test. For diagnostic purposes both should be high. The decision criterion that defines when a test is positive or negative influences the consequences of the possible diagnostic outcome. For example the physician tends to classify some normal patient as ill if the consequence is that no sick patients are misclassified. His decision also determines the cost-benefit relation for the patient.

Step 7 is hard to perform. Often the diagnostician does not know the base rates of diseases. If at all he alternatively uses rough estimates. In most cases this step, which is important for a reliable diagnosis, is simply omitted in every day practice.

Step 8 changes the pre-test probability into the post-test or a-posteriori probability (positive predictive value). The test result is a probabilistic statement.

Step 9 performs the diagnostic problem solving, i. e. the application of the nosological taxonomy for the classification of individual cases. In individual disease cases analytic and synthetic arguments which move over several levels of the decomposition-abstraction hierarchy are often difficult to justify. There are special analytical problems of diagnostic reasoning in individual cases.

Step 10 emphasizes the necessity to come to a pragmatic decision. Steps 9 and 10 express the core of diagnostic classification. Diagnostic classification is a core discipline of artificial intelligence in medicine.

Step 13 switches from serial to parallel processing of the therapeutic and prognostic implications of **PPI**. The aspects of prognostic planning are discussed below. The decision to stop the diagnostic evaluation presupposes a prognostic estimation.

Step 14 emphasizes that diagnostic tests which do not improve the prognosis of **PPI**, lead to pseudo-diagnostics. They have no therapeutic and prognostic consequences and shall be avoided. The value of a test varies with the purpose (i. e. diagnosis, monitoring, prevention) it is done for. Otherwise the principle "nil nocere" (step 1 of the algorithm) is violated.

Step 15 refers to the principle argumentative nature of diagnosis. Defending a diagnosis is in general not different from every day arguing. It is the attempt to prove the truth of a medical diagnostic statement. These arguments do not use special logical figures. Their formal structure is usually very simple: If $test_1$, $test_2$, ... , $test_n$ are correct then conclude **PPI**. In general the modified "Toulmin-Scheme"[18] is applied: From **PPPI** and **NE** follows **PPI** with probability p unless there is no exception. **PPPI** represents the individual data, **NE** represents the nosological principles, **PPI** represents the conclusion assuming there is no exception, and probability p denotes its reliability. Just like every day argumentation the limitations of such conclusions emerge from the fact that medical statements are only valid up to a certain degree of probability. Therefore probability logic must be applied to the discussion of differential diagnosis.

Step 15 makes a decision between competing diagnoses depending on their actual probability. Whether a **PPI** has reached sufficient support to be accepted as probably true depends on the predictive value of the performed medial tests and a cut-off value implemented in **NE**.

---

[18] Toulmin S. The Uses of Argument, London: Cambridge Univ. Press, 1969.

Step 17: If there is no appropriate difference between the probabilities of competing diagnoses and the same results can not rule out alternative **PPI** then the decision is impossible. In this case the diagnostician has to follow the "wait and see"-strategy: Watching **PPPI-C** and iterating the algorithm until new data support either **PPI** or alternative diagnoses.

5.3 Prognostic Planing

Rationale expectations in clinical medicine are no optimal forecasts in terms of the prognostic risk. The aim of an undistorted expectation that minimizes the prognostic risk may be realized in the case of one-dimensional patient images. If **PPI** is multi-dimensional, forecasts are much more difficult to perform. Then the solution of the individual clinical prognosis is not the inference of a pseudo-exact probabilistic statement. The individual prognosis is a genuine inexact problem which probabilistic theories can only approximately describe.

Individual prognosis in clinical medicine is based on the prognostic structuring of the individual clinical setting and on the identification of relevant clinical information which can be completed with regard to the purpose of the prognosis.

The algorithm below describes the basic heuristics of prognostic reasoning in clinical practice. Basic elements are the individual **PPI-C** and inferenced time-relative expectations.

Prognostic-planning-algorithm:
1. avoid "ud aliquit fiat";
2. identify the dimensionality of **PPI-C**;
3. decide on the independency of the involved $NE_1, \ldots NE_n$;
4. retrieve **PPI-C**$_{natural}$ for each $NE_1, \ldots NE_n$;
5. specify individual kinetics and dynamics of **PPI-C**;
6. specify the inner time structure of **PPI-C**;
7. choose therapy options for each $NE_1, \ldots NE_n$;
8. in case of emergency skip to step 11;
9. start subroutine *risk-benefit estimation*:
10. find clinical trials for $NE_1, \ldots NE_n$;
11. if 9.1 fails, go to step 9.7;
12. if the quality of data are acceptable, match **PPI-C** with the inclusion and exclusion criteria of the trials;
13. if step 9.3 fails, got to step 9.7;
14. calculate numbers needed to treat;
15. calculate numbers needed to harm;
16. compare **PPI-C** with similar cases (case reports);
17. refer **PPI-C** to subjectively observed frequencies;
18. start subroutine *cost-benefit-estimation*;
19. return to step 9.3;
20. estimate individual cost-minimization;
21. inference the expectation structures **PPI-C**$_{future}$;
22. specify the kinetics of **PPI-C**$_{future}$;
23. establish informed consent with the patient about the purpose of the prognosis;
24. initialize treatment for **PPI**;
25. if step 14 fails, wait and see;
26. specify follow-up intervals in **I**;
27. document changes in **PPI-C**;
28. estimate a prognostic error by comparing **PPI-C**$_n$ and **PPI-C**$_{n+1}$ after a follow-up interval i_{n+1};
29. if the prognostic error exceeds a cut-off value, go to step 1;
30. if **PPI** changes, go to step 1;
31. iterate algorithm.

Explanation of the algorithm: Most of the steps of the algorithm are self-explanatory, only few of them need some comment.

Step 1 refers to an old basic principle of clinicians to avoid treatment without any solid rationale. Acting solely because there has something to happen for the patient is unethical.

Step 3 is necessary to decide whether interdependencies, i.e. mutual amplifications or attenuations, between $NE_1, ..., NE_n$ may influence the individual course of **PPI-C**.

Step 4 emphasises that the physician has to know the natural course of a disease to make forecasts. The next steps of the prognostic-planning-algorithme are self-explanatory.

Step 9 refers to the evidence of treatment options. This step is a very time-consuming process and presupposes that good retrieval systems are available to use the relevant sources. If clinical trials are available, the physician has to decide whether his patient matches the inclusion or exclusion data of the study. Only in this case, study results are applicable for the individual treatment. If **PPI-C** matches the study data then the patient can be treated according to the study design. Study data are acceptable if the trial is done by real world doctors in real world settings with real world volunteers and the means by which the data have been interpreted are valid. If **PPI-C** fulfils this constraint, the next steps are step 9.5 and 9.6. They are performed to check the empirical evidence for prognostic planning by calculating the number needed to treat (NNT) and the number needed to harm (NNH). NNT and NNH are useful in conveying the results of clinical trials.[19] NNT and NNH emphasize the effort and the risk to reach a single tangible result. NNT is a measure for the efficiency of treatment and expresses the benefit of an active treatment over a control. NNT is the number of patients who need to be treated to prevent one adverse outcome. It is the quotient of 1 divided by the absolute risk reduction (ARR) (NNT = 1/ARR) accompanied by its 95% confidence inter-

---

[19] Laupacis A., Saenett DL, Roberts RS. An Assessment of clinically useful measures of the consequences of treatment. N Engl J Med 318; 1988: 1728-33. Sackett DL, Haynes RB. Summarizing the effects of therapy: a new table and some more terms. Evidence-Based Medicine 2, 1997, 103.

val. Absolute risk reduction, relative risk, and odd ratio are basic measures of association. In the same manner the frequency of side effects of a treatment are expressed by NNH. The physician compares NNT and NNH to determine the therapeutic efficiency of an intervention.

Two examples can illustrate the advantages of these simple measures of frequencies. For oral anticoagulation of patients with arterial fibrillation, a very common heart rhythm disturbance, the NNT is 12 and the NNH 52. This means that after a year-long treatment of twelve patients one apoplectic stroke has been prevented. During the same time period one patient out of 52 suffers a cerebral haemorrhage as complication of the oral anticoagulation. Four prevented strokes are the advantage over one cerebral haemorrhage by the treatment. If the intensity of anticoagulation is reduced NNT for prevented strokes of patients older than 65 years is 15-25, whereas NNH rises to 500 for minor bleedings and 330 for major bleeding complications. The benefit of a therapy is small if the difference between NNT and NNH is small. The NNT for antibiotic therapy of otitis media is 17. The NNH is 18, that is one of 18 children suffers severe complications of antibiotic therapy.

Often there are no appropriate clinical trials to support clinical decision making. If **PPI-C** is multi-dimensional or if **PPI** represents an infrequent disturbance, statistical evidence is usually missing.

Mesenteric vein thrombosis is an example of such a rare but dangerous cause for mesenteric ischemia. In case of disorders of low incidence treatment options are based on clinical experience together with pathophysiologiocal understanding of the disease process. Randomized trials comparing various therapeutic strategies are hardly to be expected. Then predictions depend on case reports (step 9.7) and knowledge about the individual course of **PPI-C**. In this case clinical arguments are based on analogies. Analogies are logically not stringent, because the general principle or statement is missing which would justify the passage from one case to the other.

Step 9.7 is usually combined with step 9.8. If general knowledge is missing forecasts about **PPI-C** are additionally based on personal expert knowledge of the clinician and his colleagues (peer group).

Step 9.8 permits that the truth of a clinical forecast depends on the authority of a peer group or a single member, a single clinician who makes the statement. The truth of the argument depends on its source itself. This approach presupposes the principle that all statements of the source are true. This type of argumentation is still popular among physicians. On account of missing empirical data and for the sake of convenience the source of the assertion is substituted for its rationale verifiability.

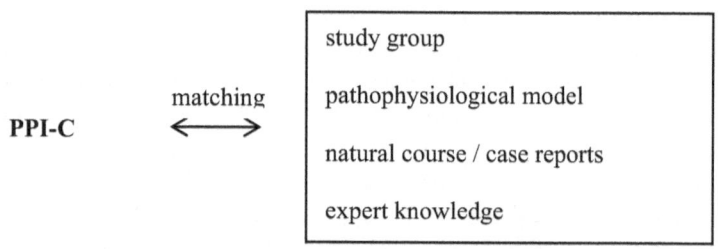

Figure 9. Summary of the processing moduls of step 9 of the prognostic-planing algorithm

Step 10 of the prognostic-planning algorithm refers to the cost-benefit relation for the individual patient. Usually cost-effectiveness analysis is understood as efficiency of resource allocation in relation to the number of successfully treated patients. Here, cost-benefit analysis is individualized. The estimation refers to the treatment cost for the individual treatment.

Step 11 performs the essential prognostic inference of **PPI-C**$_{future}$. That is the projection of **PPI-C** into the future (figure 8). In figure 8 **PPPI-C** changes to **PPI-C** after **PPI** has been identified.

Step 13 demands that the personal value system of **PSI** and **DSI** has to be matched. Subjective feelings and the patient's subjective functional disability have to be correlated with objective disease criteria. The quality of life of patients with chronic diseases is an important criterion for medical forecasts.

Step 14 refers to the clinical setting that no rationale and appropriate treatment is available. Then the clinician has to follow the "wait-and-see" strategy. The next steps of the algorithm lead to a documented course of **PPI** and to an estimation of the prognostic error. The prognostic error is necessary to plan corrections of the disease management of the patient.

Step 19 determines an arbitrary cut-off value up to which the prognostic error is tolerable. This cut-off is a value which has to be elaborated step by step in the medical management process. There are no general empirical data for its definition. The clinician has to decide in cooperation with the patient and according to the actual individual clinical situation of **PPI-C**.

Step 20 is necessary if **PPI** changes. Then the algorithm starts again at once.

Step 21 expresses the fact that the prognostic-planning algorithm continuously runs parallel to the other two algorithms during the disease management process.

5.4 Argumentation Space

The history-based-learning algorithm (HBL-algorithm), the diagnostic-inferencing algorithm (DI-algorithm) and the prognostic-planing algorithm (PP-algorithm) work sequentially and parallel (figure 10).

Uterrances of the patient and events are the input for the history-based- learning algorithm. HBL produces **PPPI**. If HBL fails, it starts again or it stops, if no medical problem has been identified. **PPPI** is the input for the diagnostic inferencing algorithm. Its result is **PPI**. If **PPI** needs no treatment and is spontaneously reversible according to the natural course describes in **NE**, processing stops at this point. In all other cases **PPI** is the input for the prognostic planning algorithm. Simulatneously **PPI** is refined and defended against differential diagnosis by entering the feed-back loop. While **PPI** is refined the PP-algorithm produces **PPI-C**. If **PPI-C** is spontaneously reversible and a treatment plan is generated, processing stops at this point. If **PPI-C** is spontaneously irreversible or progressive, **PPI-C** enters the feedback loop. The time course of the feed-back loop depends on the kinetics of the diseases describes in **NE**. If the sequence fails to

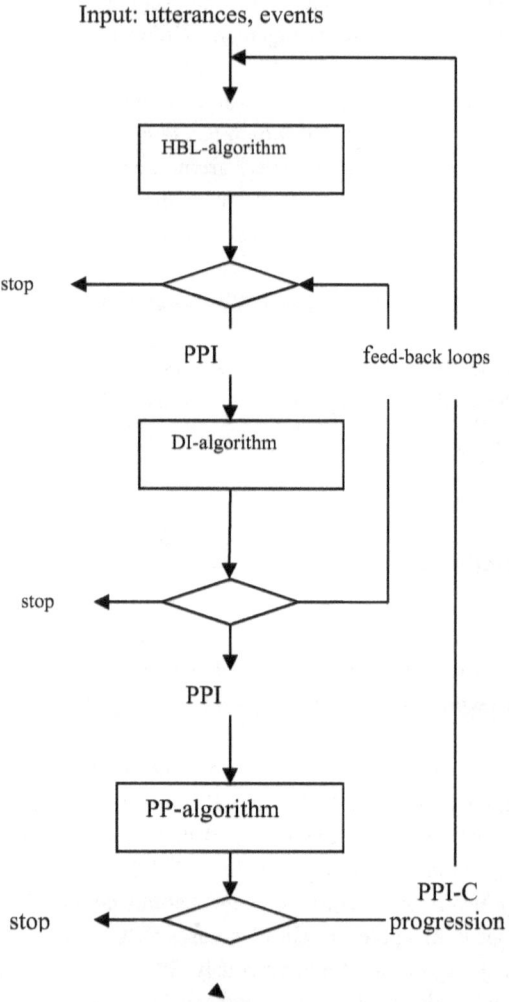

Figure 10: Parallel processes. HLB = history-baes learning, DI = diagnostic inferencing, PP = prognostic planning. Further explanations in the text.

produce **PPI**, **PPPI-C** is used as default for the prognostic planning algorithm. The situation after three runs of the HBL-DI-PP sequence of algorithms the situation can be presented by AD-tables (figure 11). The AD-tables in figure 8 show the data for PPI at the time point t_1, t_2 and t_3 after three runs. Their sequence defines the argumentation space for **PPI-C** or **PPPI-C**, if the algorithms fail to produce a believable **PPI**.

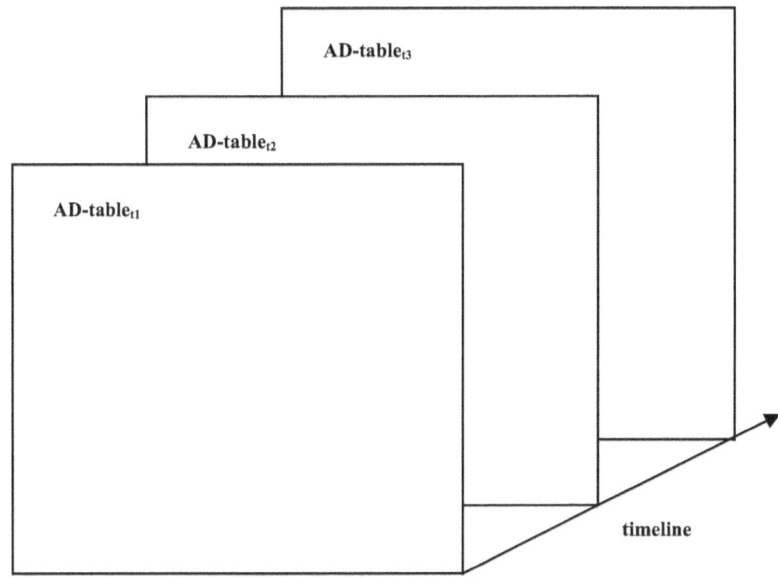

Figure 11: Argumentation space for **PPI-C** after three runs of the algorithms represented as AD-tables over the time.

6. Summary

The proposed analysis of the logical structure of the practice of medicine started with three axioms: (1) the axiom of iconographic reasoning in clinical medicine, (2) axiom of interacting self-images, and (3) the axiom of data approximation. The first stands for that the physician uses incomplete and fuzzy clinical data sets to create a pattern or image of the ill subject in its individual domain. The second presupposes that in clinical medicine all interactions are deeply rooted in the patient's self-image and in the doctor's self-image. The third axiom denotes that images of a patient in the sense of medical data representations are always only approximations of the true situation.

To stress the iconographic view in clinical medicine the abstract terms **PSI** (patient self image), **PPPI** (partial potential patient image) and **PPI** (potential patient image) are introduced which describe the image-concepts for practical clinical work on a very general formal level. The "image"-term stresses the so called iconographic view of the physician on his patient.

PSI is composed of four contexts. The organic and psychic contexts represent the "internal" world of a patient, the socio-cultural and environmental contexts constitute his "external" world. The patient subjectively interprets (interpretation function δ) the different domains of his internal and external world and creates an image of his own, his self-image. The same happens on the doctor's side and constitutes **DSI**. Professional medical knowledge and technical medical skills are the characteristic parts of **DSI**.

A decomposition hierarchy and an abstraction hierarchy for medical descriptions specify the features of the medical knowledge. Both hierarchies form the nosological object matrix for nosological nets (**NN**) whose generic units are disease entities, in my analysis referred to as nosology entities (**NE**). **NE** denotes an autonomous knowledge unit that can be used to formulate reasonable diagnostic and therapeutic propositions. Nosology entities consist of nosology concepts (**NC**) and intended medical applications (**IMA**). **NC** are theoretical descriptions of diseases. Intended medical applications are typical clinical examples of a disease which illustrate the range and the meaning of a nosology concept.

After clarifying the structures of medical knowledge patient images are classified. A patient image is seen as a true copy of **PSI**. True patient images (**PI**) are an ideal pretending the direction and the aim of a diagnostic process. Good approximations of **PI** are partial potential patient images (**PPPI**) and potential patient images (**PPI**). Their difference can be compared with the difference between the "presumptive diagnosis" and the "established diagnosis" in common clinical language. The nosological classification function θ in **PPI** makes the difference.

Dynamics and kinetics of both, **PPPI** and **PPI**, are described in courses. The disease course is a very important and basic source for medical reasoning. The inner time structure of the course of **PPI** (**PPI-C**) is **PPI-C**$_{latency}$ < **PPI-C**$_{onset}$ < **PPI-C**$_{duration}$. The natural course (**PPI-C**$_{natural}$) of a disease is its course without medical interventions. **PPI-C**$_{natural}$ has to be considered to come to a rational prognosis.

Prognosis is the most interesting aspect of **PPI-C**$_{natural}$. I argue that the expected future course is a conditional prognosis. It is a projection of the history of **PPI-C** into the future presupposing that the future events in **PPI-C** are structurally similar to the past **PPI-C**, in other words the expectation structure of the future **PPI-C** is homomorphic to the event structure of the past **PPI-C** (*Homomorphism axiom of prognosis*). Forecasts are based on the individual **PPI-C**. Often, predictions convert from rational **NE**-based prognoses to estimates about future trends of individual clinical situations.
The heuristics for processing the introduced objects are then composed into three algorithms which reflect three basic cognitive tasks during the practice of medicine: history-based learning, diagnostic inferencing, and prognostic planning.

History-based learning is the initial task of the patient-physician interaction. It can be reconstructed as cross domain learning by task embedding. Its result is **PPPI**. Diagnostic reasoning aims at the production of **PPI**. Both **PPPI** and **PPI** incorporate historical aspects of the relevant medical events. In contrast to the historical view prognostic planning considers the future development of **PPI** or **PPPI**. Rationale expectations in clinical medicine are no optimal forecasts in terms of the prognostic risk. The individual prognosis is a genuine inexact problem which probabilistic theories can only approximately describe. Individual prognosis in clinical medicine is based on the prognostic structuring of the individual clinical setting and on the identification of relevant clinical information which can be completed with regard to the purpose of the prognosis. The prognostic planning algorithm describes the basic heuristics for prognostic reasoning in clinical practice.

7. Exercises: Tools for Cognitive Processing

In 1988 J. Willis Hurst outlined the needs for helps for the practicing physician: „One needs tools with which to think. Once the tools for thinking have been identified, it is necessary to use them properly and frequently in order to improve thinking. Use of the tools for thinking can then be viewed as thinking skills. Such skills must be understood and perfected just as one perfect any other skills. "[20]

Implications of the proposed theory for cognitive engineering are discussed. They formulate exercises to construct such "tools for thinking" in clinical medicine.

We already have several existing applications for the automatic interpretation of medical data and data monitoring at the D4A2-level of the decomposition abstraction hierarchy. They represent different aspects of **PPPI** or **PPI**. However, they do not provide a holistic view on the interconnected health problems of a patient. Examples are automatic interpretation of e.c.g. (electro-cardiogram) or lung function data interpretation. They are judgment helps at the organ function level. Automatic data interpretations at this level of abstraction presuppose exactly defined contexts which are suitable for simple classifications. Typically, well defined medical data are interpreted by a couple of simple rules. Other examples of the D8A2-, D9A2-, or D10A2-level are routine laboratory monitoring systems which indicate contraindications of treatment, missing data for planed therapies. Other programs support pharmacotherapy or examine interactions between different drug therapies. They are useful management helps for the physician and may be very effective and even cost efficient if integrated into clinical or office information systems.

If we look on aspects of heuristic processing, computer support for the production of **PPPI**, **PPI**, **PPPI-C** and **PPI-C** is more complex. Such programs work on the D1A4 or D1A5 level of the abstraction hierarchy. In conclusion

---

[20] Hurst, J.W. (ed.), a.a.O., p 13.

I outline possible performances of useful tools on higher levels of abstraction for cognitive processing helps in practical medicine.

History taking: For the production of **PPPI** the practicing physician needs a documentation system for verbal and non verbal utterances of his patient. A tool for history taking supports the generation of a problem list PL_{pat}. The tool makes proposals for further evaluations of the patient's history with regard to the actual PL_{pat}. The tool supports case based comparisons between old histories and the new patient history by matching $PPPI_{new}$ with $PPPI_{old}$.

Holistic view: If $PPPI_{new}$ is unchanged over a defined period of time, a tool generates a synopsis of the patient's world. The switch from the analytical investigation back to the holistic view of the patient's world including his wishes and his intentions is essential for the appropriate management of PL_{pat} (compare step 5 of the **PPPI**-learning-algorithm).

Transforming **PPPI** into **PPI** is the diagnostic step. Usually clinical situations are ambiguous and risky. To minimize the risk of clinical decisions a tool for cognitive processing has to support several steps of the diagnostic-inferencing-algorithm.

Diagnostic test selection: The tool has to propose diagnostic tests for **PPPI** according to their power to detect strong disease indicators. The tool references the specificity and sensitivity of the chosen diagnostic tests. The system proposes predictive values of tests ordered by the physician. The system automatically checks whether the predictive value of an ordered test is acceptable according to the actual available medical data. If not the system proposes an alternative test whose predictive value is higher according to the actual diagnostic setting and already available data.

Local base rates: The data documentation system produces base rate estimations of the diseases represented in the local information system. The predictive value of tests is useless if the base rates of the tested diseases are unknown. The tool documents and retrieves age-dependent disease base rates which refer to the actual population the physician has to take care of.

Knowledge retrieval: An important task is the retrieval of new knowledge being relevant for the individual medical case. The program automatically generates a trace through different knowledge bases and retrieves knowledge that helps to avoid argumentation fractures and loops. An important function is the automatic electronic record based examination whether the individual patient matches the inclusion and exclusion criteria of clinical studies being relevant for his treatment. This is an example of a simple but very effective help for the physician. The manual retrieval of scientific data during the everyday routine work is difficult to manage.

Decision strategies: Physicians have sometimes to change search strategies to solve their diagnostic or therapeutic problems. Then, it may be necessary to change from topographical search to a "hypothesize-and-test"-strategy to enhance processing capacity and search efficiency. The change of the problem solving strategy in this situation may lead to a more effective processing of causal relations in nosology nets (**NN**). A "tool for thinking" recommends a search strategy which is more suitable in the individual clinical situation. The role of decision strategies in ambiguous and risky situations of everyday life has been investigated by M. Brand and co-workers. They found that calculative processes and strategies improve decision-making under risk conditions. "In contrast, individuals who decide intuitively, even in rather simple decisions between only two alternatives, tend to risky and disadvantageous choices in a complex task measuring decision under risk conditions".[21]

Argumentation check: This tool records the arguments used in the heuristic algorithms for the construction of **PPPI** and **PPI**. Such a tool supports an argumentation control. The program examines the completeness of the argumentation whether there are mistakes or jumps in it. Arguments use general principles which can be grouped into argumentation schemes.[22] If causal argumentations move bottom-up in the decomposition hierarchy or from left to the right in the abstraction hierarchy the program may detect short cuts and

---

[21] Brand M, Heinze K, Labudda K, Markowitsch HJ. The role of strategies in deciding advantageously in ambiguous and risky situations, Cogn Process 2008; 9: 159-173, p 168.

[22] Rahwan I, Simari GR (eds.), Argumentation in Artificial Intelligence. Berlin, Heidelberg, New York: Springer, 2009.

fallacies. The program has to answer questions like: "Which arguments let me to my last medical decision concerning PPI_{new}?" It would be of help, if the program visualizes the movements of the arguments through the argumentation space. The matrix of the argumentation space is the decomposition-abstraction hierarchy. In analogy to a magnetic resonance tomograph a knowledge- based tomograph produces images of our incomplete causal argumentation space. The physician uses this image to reflect on his arguments to avoid fallacies. The program may perform a consistency check of his argumentation.

*Feedback by Interacting **DSI***: The program records how the physician usually decides, how he uses problem solving strategies. These observations are used by the program to reconstruct a personal intra-individual decision model of the physician. If several physicians check their decision behaviour by the program, the interaction of **DSI** leads to an incremental process-feedback. Such interactions of physicians working together in a common medical discipline have been studied by J. R. Kirwan.[23] Their findings show that a process oriented continuous feedback leads to better results of a group of specialists. Feedback leads to a better match of their decision criteria, avoids discrepancies in their estimates of disease stages, improves consistent therapeutic consequences. A tool for supporting cognitive processing should embody such a feedback function based on protocols of the argumentations and the decision behaviours of its user.

*Visualizing **PPI-C***: To inference an expectation structure is the core activity of the prognostic-planning-algorithm. Each step of the algorithm can be seen as programming task. A visual representation of **PPI-C**, however, would be most suitable to improve individual forecasts. These visualizations summarize the available patient data and provide an impression how **PPI** will progress under the influence of interacting disease processes. It is this feeling for the kinetics and dynamics of the individual course that guides the experienced physician to a prognostic estimate.

---

[23] Kirwan J.R. et al, Analysis of clinical judgment helps to improve agreement in the assessment of rheumatoid arthritis, Annals of the Rheumatic Diseases 47, 1988, 138-143.

8 Appendix

8.1 Sets

$BS_1, ..., BS_n$:	not empty basic sets for **NE**
$\{cm_1, ..., cm_n\}$:	set of clinical manifestations
C_{org}:	set of organic contexts
C_{psy}:	set of psychological contexts
C_{soc}:	set of socio-cultural contexts
C_{env}:	set of environmental contests
$\{dm_1, ..., dm_n\}$:	set diagnostic manifestations
DSI:	set of doctor self-images
IMA	set of intended medical applications
NE:	set of nosology elements
NN:	set of nosology nets
NV:	set of non-verbal utterances
PI:	set of patient images
PPI:	set of potential patient images
PPI-C	set of courses of **PPI**
PPPI:	set of partial potential patient images
PSI:	set of patient self-images
V:	set of verbal utterances

8.2 Relations and Functions

Relations
$S \subseteq \text{Pot}(\mathbf{NE}) \times \text{Pot}(\mathbf{NE})$ is a specialisation relation.
$<$ is an antesessor relation.

Functions
$v: \mathbf{C}_{org} \times \mathbf{C}_{psy} \times \mathbf{C}_{soc} \times \mathbf{C}_{env} \rightarrow \mathbf{V}_{pat}.$
$nv: \mathbf{C}_{org} \times \mathbf{C}_{psy} \times \mathbf{C}_{soc} \times \mathbf{C}_{env} \rightarrow \mathbf{NV}_{pat}.$
$\delta: \mathbf{V} \times \mathbf{NV} \rightarrow \{\text{healthy, indifferent, sick}\}.$
$\sigma: \mathbf{DSI} \times \mathbf{V}_{pat} \times \mathbf{NV}_{pat} \rightarrow \mathbf{PL}_{pat},$
$\pi: \mathbf{DSI} \times \mathbf{PL}_{pat} \rightarrow \mathbf{IP}_{pat}.$
$\theta: \mathbf{PPPI} \rightarrow \mathbf{NN}$

8.3 Definitions

Definition 1: PSI is a set of patient self-images iff for C_{org}, C_{psy}, C_{soc}, C_{env}, V_{pat}, NV_{pat} here are the functions v, nv, δ such that:
1. **PSI** = $\langle C_{org}, C_{psy}, C_{soc}, C_{env}, V_{pat}, NV_{pat}\ v, nv, \delta \rangle$,
2. C_{org} is a set of organic contexts,
3. C_{psy} is a set of psychological contexts,
4. C_{soc} is a set of socio-cultural contexts,
5. C_{env} is a set of environmental contests,
6. v: $C_{org} \times C_{psy} \times C_{soc} \times C_{env} \to V_{pat}$,
7. nv: $C_{org} \times C_{psy} \times C_{soc} \times C_{env} \to NV_{pat}$,
8. for all C_{org}, C_{psy}, C_{soc}, C_{env}:
δ: $V_{pat} \times NV_{pat} \to$ {healthy, indifferent, sick}.

Definition 2: DSI is a set of doctor self-images iff for C_{org}, C_{psy}, C_{soc}, C_{env}, V_{doc}, NV_{doc} there are the functions v, nv, δ such that:
1. **DSI** = $\langle C_{org}, C_{psy}, C_{soc}, C_{env}, V_{doc}, NV_{doc}\ v, nv, \delta \rangle$,
2. C_{org} is a set of organic contexts,
3. C_{psy} is a set of psychological contexts,
4. C_{soc} is a set of socio-cultural contexts,
5. C_{env} is a set of environmental contests,
6. v: $C_{org} \times C_{psy} \times C_{soc} \times C_{env} \to V_{doc}$,
7. nv: $C_{org} \times C_{psy} \times C_{soc} \times C_{env} \to NV_{doc}$,
8. for all C_{org}, C_{psy}, C_{soc}, C_{env}:
   δ: $V_{doc} \times NV_{doc} \to$ {healthy, indifferent, sick}.

Definition 3: NE is a nosology entity iff there are a nosology concept **NC** and a set of intended applications **PPI** such that
NE = $\langle$**NC, PPI**$\rangle$.

Definition 4: **IMA** are intended application of **NE** iff there are BS_{NE}, M_{NE}, DM_{NE} so that:
1. **IMA** = < BS_{NE}, M_{NE}, DM_{NE} >;
2. BS_{NE} is a not empty basic set;
3. M_{NE} is a set of manifestations;
4. DM_{NE} is a set diagnostic manifestations,
5. $DM_{NE} \subseteq M_{NE}$;
6. **IMA** performs the causal rules of NC_{NE}.

Definition 5: An attribute of an element of the basic set BS_{NE} is a diagnostic manifestation dm $\in DM_{NE}$ with reference to **NE** iff each determination of dm in an application of **NE** presupposes at least one concrete intended medical application *ima* $\in$ **IMA** of **NE**.

Definition 6: **NN** is a nosology net iff
1. **NN** = < Pot (**NE**), S>
2. Pot (**NE**) is a non empty set
3. S $\subseteq$ Pot (**NE**) × Pot (**NE**) is a specialisation relation.

Definition 7: **PPPI** is a set of partial potential patient images iff for **DSI**, V_{pat}, NV_{pat}, **PL** there are the functions σ and π such that:
1. **PPPI** = < **DSI**, V_{pat}, NV_{pat}, PL_{pat}, IP_{pat}, σ, π>,
2. **DSI** is a set of doctor self-images,
3. σ: $DSI \times V_{pat} \times NV_{pat} \to PL_{pat}$,
4. π: $DSI \times PL_{pat} \to IP_{pat}$.

Definition 8: **PPI** is a set of potential patient images iff for **PPPI**, **NN** there is the function θ such that:
1. **PPI** = < **PPPI**, **NN**, θ>,
2. θ: $NN \times PPPI \to PPI$.

Definition 9: **PPI-C** is a course of **PPI** iff there are **PPI**, **I** and < such that
1. **PPI-C** = ⟨**PPI**, < , **I**⟩,
2. **I** is a non-empty set of time intervals,
3. < is an antesessor relation in **PPI**.

Definition 10: **PPPI-C** is a course of **PPPI** iff there are **PPI**, **PPPI**, **I** and < such that
1. **PPI** is empty
2. **PPPI-C** = ⟨**PPPI**, < , **I**⟩,
3. **I** is a non-empty set of time intervals,
4. < is an antesessor relation in **PPPI**.

8.4 Algorithms

PPPI-learning-algorithm:
1. start patient-physician communication;
2. list the non-verbal NV_{pat}
3. list verbal utterances V_{pat};
4. for NV_{pat} and V_{pat} find the corresponding patient contexts, C_{org}, C_{psy}, C_{soc}, C_{env};
5. match $\langle C_{org}, C_{psy}, C_{soc}, C_{env}, V_{pat}, NV_{pat} \rangle$ with the understanding and intention of the patient;
6. $\sigma: DSI \times V_{pat} \times NV_{pat} \to PL_{pat}$;
7. $\pi: DSI \times PL_{pat} \to IP_{pat}$;
8. produce $PPPI_{new}$;
9. elaborate the history of $PPPI_{new}$;
10. match $PPPI_{new}$ with $PPPI_{old}$;
11. iterate algorithm until $PPPI_{new}$ is unchanged over a defined time period $[t_1, \ldots, t_n]$.

Diagnostic reasoning-algorithm

1   respect "nil nocere";
2   start with **PPPI$_{new}$**;
3   select diagnostic tests according to their power to detect strong disease indikators, consider existing reasoning guidelines;
4   reference the specificity and sensitivity of the chosen diagnostic tests;
5   perform the tests sequentially or parallel depending on the acuity of the problem;
6   reference the disease base rates detected by the corresponding tests, if unknown, estimate them;
7   evaluate the pre-test and post-test probabilities of test results;
8   perform nosological classification function θ;
9   make a diagnostic decision and define **PPI$_{new}$**, if impossible, iterate algorithm until **PPI$_{new}$** is believable;
10   start prognostic planning algorithm;
11   refine **PPI$_{new}$** as long as its prognosis improves;
12   defend **PPI$_{new}$** against differential diagnosis, if impossible, iterate algorithm until **PPI$_{new}$** can be defended, if not possible, watch **PPI-C**.

Prognostic planning-algorithm
1. avoid "ud aliquit fiat"
2. identify the dimensionality of **PPI-C**;
3. decide on the independency of the involved $NE_1, \ldots NE_n$;
4. retrieve **PPI-C**$_{natural}$ for each $NE_1, \ldots NE_n$;
5. specify individual kinetics and dynamics of **PPI-C**;
6. specify the inner time structure of **PPI-C**;
7. choose therapy options for each $NE_1, \ldots NE_n$;
8. in case of emergency skip to step 11;
9. start subroutine *risk-benefit-estimation*:

9.1 find clinical trials for $NE_1, \ldots NE_n$;
9.2 if 9.1 fails, go to step 9.7;
9.3 if the quality of data are acceptable, match **PPI-C** with the inclusion and exclusion criteria of the trials;
9.4 if step 9.3 fails, got to step 9.7;
9.5 calculate numbers needed to treat;
9.6 calculate numbers needed to harm;
9.7 compare **PPI-C** with similar cases (case reports);
9.8 refer **PPI-C** to subjectively observed frequencies;
10. start subroutine *cost-benefit-estimation*;
10.2 return to step 9.3;
10.3 estimate cost-minimization;
11. inference the expectation structures **PPI-C**$_{future}$;
12. specify the kinetics of **PPI-C**$_{future}$;
13. establish informed consent with the patient about the purpose of the prognosis;
14. initialize treatment for **PPI**;
15. if step 14 fails, wait and see;
16. specify follow-up intervals in **I**;
17. document changes in **PPI-C**;
18. estimate a prognostic error by comparing **PPI-C**$_n$ and **PPI-C**$_{n+1}$ after a follow-up interval i_{n+1};
19. if the prognostic error exceeds a cut-off value, go to step 1;
20. if **PPI** changes, go to step 1;
21. iterate algorithm.

9 References

Adler J, Interlandi J. The hospital that could cure health care. Cleveland Clinic is both highly effective and fiercely efficient. So why are its methods so rare? Newsweek, December 7 2009, 44 - 47.

Allen JF. Towards a general theory of action and time, Artificial Intelligence, 23, 123-154, 1984.

Allen JF, Hayes PJ. Moments and points in an interval-based temporal logic, Computational Intelligence, 5, 225-238, 1989.

Altman, DG. Practical Statistics for Medical Research. Boca Raton, London, New York, Washington DC, Chapman & Hall/CRC 1999

Beek P van. Reasoning about qualitative temporal information, Artificial Intelligence, 58, 297-326, 1992.

Balzer, W., Moulines C. U., Sneed, J.D.: An Architecture for Science, The Structuralistic Program, Dordrecht: D. Reidel Publ., 1987.

Benthem JFAK van. The Logic of Time, An Model-Theoretic Investigation into the Varieties of Temporal Ontology and Temporal Discourse, Dordrecht: Reidel Publ., 2nd Edition, 1992.

Blum RL. Discovery and Representation of Causal Relationships from a Large Time-Oriented Clinical Database: The RX Project, Lecture Notes in Medical Informatics 19, Berlin, Heidelberg, New York: Springer Verlag, 1982

Blois MS. Information and Medicine, The Nature of Medical Descriptions, Berkeley: Univ. of California Press, 1984.

Bouzid M, Ligeza A. Temporal Logic based on characteristic functions, in: Wachsmuth I, Rollinger C-R, Brauer W (eds.). KI-95: Advances in Artificial Intelligence, Lecture Notes in Artificial Intelligence 981, Berlin, Heidelberg: Springer Verlag, 1995, 221-232.

Brand M, Heinze K, Labudda K, Markowitsch HJ. The role of strategies in deciding advantageously in ambiguous and risky situations, Cogn Process 2008; 9: 159-173.

Braunwald E, Isselbacher KJ, Petersdorf RG, Wilson JD, Martin JB, Fauci AS (eds.). Harrison´s Principles of Internal Medicine, New York, et al.: McGraw-Hill Book Company, 11th ed. 1987.

Charles ED, Dustin LB, Hepatitis C virus-induced cryoglobulinemia, Kidney international 76: 2009, 818-824.

Cummins R. Representation, Targets, and Attitudes, Cambridge Mass: MIT Press, 1996.

Del Mar C. u. Mitarb., Brit. Med. J., 314, 1997, 1526.

Gigerenzer G. Gut Feelings. New York: Viking. 2007

Gigerenzer G. Fast and Frugal heuristics: tools of bounded rationality, in: Koehler D, Harvey N (eds.): Blackwell handbook of judgement and decision making. Oxford 2004, 62-88.

Heller B, Meyer-Fujara J, Schlegelmilch S, Wachsmuth I. HYPERCON: ein Konsultationssystem zur Hypertonie auf der Basis modular organisierter Wissensbestände. University of Bielefeld ; Fakultäten. Technische Fakultät, in: G. Barth et al.: Anwendungen der künstlichen Intelligenz: KI-94, 18. Fachtagung für Künstliche Intelligenz, Saarbrücken, 22./23. September 1994, Berlin: Springer: 1994, 155-169.

Heller B. Modularisierung und Fokussierung erweiterbarer komplexer Wissensbasen auf der Basis von Kompetenzeinheiten. Bielefeld, Univ., Diss., 1995. Sankt Augustin: Reihe: Dissertationen zur künstlichen Intelligenz. 1996.

Heller B, Herre H, Lippoldt K, Loeffler M. Standardized Terminology for Clinical Trial Protocols Based on Top-Level Ontological Categories. In: Kaiser K et al (eds.) Computer-based Support for Clinical Guidelines and Protocols, IOS Press 2004, 46-60.

Higgins JPT, Green S (eds.). Cochrane Handbook for Systematic Reviews of Interventions (Version 5.0.1). The Cochrane Collaboration 2008. www.cochrane-handbook.org

Hucklenbroich P. Steps towards a theory of medical practice, Theor Med. Bioeth. 19, 1998: 215-38.

Hurst, J.W. (ed.), Medicine for the Practicing Physician, 2nd edition, Boston et. al.: Butterworth, 1988.

Kahneman D, Slovic P, Tversky A (eds.). Judgement under Uncertainty: Heuristics and Biases. Cambridge, England: Cambridge University Press 1982

Knapp H. G. Logic of Prognosis (in german: Logik der Prognose, Semantische Grundlagen technologischer und sozialwissen-schaftlicher Vorhersagen), Freiburg/München, Verlag Karl Alber, 1978.

Kamp J.A.W., Tens Logic and the Theory of Linear Order, PH.D.Thesis, University of California, Los Angeles 1968

Kirwan J.R. et al, Analysis of clinical judgment helps to improve agreement in the assessment of rheumatoid arthritis, Annals of the Rheumatic Diseases 47, 1988, 138-143.

Laupacis A., Saenett DL, Roberts RS. An Assessment of clinically useful measures of the consequences of treatment. N Engl J Med 318; 1988: 1728-33

Macleod John (ed.). Clinical Examination. Edingburgh, London, NewYork: Churchll Livingstone, 1979.

Meyer-Fujara J, Heller B, Schlegelmilch S, Wachsmuth I. Knowledge-level modularization of a complex knowledge base, in: Bernhard Nebel et al. Advances in artificial intelligence: proceedings, eds. (Lecture notes in computer science; 861), Berlin: Springer, 1994, 214-225.

Müller U. SCREEKON: Medizinisches Konsultations-system für Screeningfunktionen am Beispiel des therapeutischen Managements von Patienten mit cerebralen Durchblutungsstörungen, Angewandte Inforamtik 2/1989, 76-

Müller-Kolck U. Medizinische Therapieentscheidungen mit SCREECON, in: Savory SE (ed.) Expertensysteme: Nutzen für ihr Unternehmen. Ein Leitfaden für Entscheidungsträger. München, Wien: R. Oldenbourg; 1989.

Müller-Kolck U. Expert system support for the therapeutical management of cerebrovascular disease. Artificial Intelligence in Medicine, 2: 1990, 35-42.

Müller-Kolck U. Diagnostic Consultations: A Speech-Act Theoretical Reconstruction fort he Design of Consultations Systems. Methods of Information, 1991, 311-315.

Müller-Kolck U. Expertensysteme als metadiagnostische Hilfsmittel in ärztlichen Entscheidungsprozessen, in: Hucklenbroich P, Toellner R (eds.) Künstliche Intelligenz in der Medizin. Klinisch-methodologische Aspekte medizinischer Expertensysteme, Stuttgart, Jena, New York: Fischer Verlag, 1993, 141-159.

Müller-Kolck U. Basic structures of nosology in medical arguments (in german), in: Meggle G., Nida-Rümelin J., Perspektiven der Analytischen Philosophie, Band 18, Proceedings of the 2nd Conference „Perspectives in Analytical Philosophy", Vol III, Berlin, New York: De Gruyter 1997, 518-528.

Müller-Kolck U. Modellierung individueller Prognosen in der klinischen Medizin. Theory Biosci, 120: 2001, 45-56.

Murphy EA. Probability in Medicine. Baltimore, London: The Johns Hopkins University Press, 1979.

Murphy EA. Skepsis, Dogma and Belief. Uses and Abuses in Medicine. Baltimore, London: The Johns Hopkins University Press, 1981.

Murphy EA. The Logic of Medicine, Sec Ed., Baltimore, London: The Johns Hopkins University Press, 1997.

Rahwan I, Simari GR (eds.), Argumentation in Artificial Intelligence. Berlin, Heidelberg, New York: Springer, 2009.

Rasmussen J. Information Processing and Human-Machine Interaction: An Approach to Cognitive Engineering, New York, Amsterdam: Elsevier, 1986.

Rasmussen J. Modelling distributed decision making, in: Rasmussen J, Brehmer B, Leplat J (eds.): Distributed Decision Making: Cognitive Models for Cooperative Work, John Wiley 1991, 111-142.

Richter K, Lange S. Methoden der Diagnoseevaluierung, Internist 1997, 38, 325-336

Sadegh-Zadeh K, The notion of disease and nosological system (in german, Metamed 1, 1977, 4-41.

Sadegh-Zadeh K. Foundations of clinical praxiology, part II: categorial and conjectural diagnosis, Metamedicine 1982: 159-191.

Sadegh-Zadeh K. Fundamentals of clinical methodology: 1. Differential indications, Artificial Intelligence in Medicine 6, 1994, 83-102

Sadegh-Zadeh K. Fundamentals of clinical methodology: 2. Etiology, Artif. Intell Med 1998: 227-270.

Sackett DL, Haynes RB. Summarizing the effects of therapy: a new table and some more terms. Evidence-Based Medicine 2, 1997, 103.

Schwartz S, Griffin T. Medical Thinking. The Psychology of Medical Judgment and Decision Making, Springer: New York, 1986.

Sneed JD, The Logical Structure of Mathematical Physics, Dordrecht: Reidel Publ. Co., sec. edition 1979.

Suppes P., Introduction to Logic. New York, Cincinnati, Toronto, London, Melbourne: Van Nostrand Reinhold Company, 1957.

Toulmin S. The Uses of Argument, London: Cambridge Univ. Press, 1969.

Wachsmuth I, Myer-Fujara J. Addressing the retrieval problem in large knowledge bases. Universität Bielefeld; Technische Fakultät. MOSYS-Report (3), 1990, 1-13.

Weed, L.L.: Medical Records, Medical Education and Patient Care – The Problem-oriented Record as a Basic Tool. Cleveland, Ohio: Press of Case, Western Revers University, 1970.

Weed, L.L.: Knowledge Coupling; New Premises and New Tools for Medical Care and Education, Berlin et. al.:Springer-Verlag, 1991.

About the Author
The author studied medicine and philosophy of science at the Universities of Marburg and Münster (Germany), medical degree (MD) in 1983 from the University of Münster, Dr med in 1985 for research in theoretical medicine, and 1989 diploma in information science from the University of Konstanz (Germany). He is a practicing physician since 1983 and specialized in internal medicine, diabetology, angiology, and hemostaseology.

www.ingramcontent.com/pod-product-compliance
Lightning Source LLC
Chambersburg PA
CBHW030448220526
45464CB00006B/2448